Urban and Regional Technology Planning

Planning Practice in the Global Knowledge Economy

Kenneth E. Corey and Mark I. Wilson

LONDON AND NEW YORK

First published 2006
by Routledge
2 Park Square, Milton Park, Abingdon, Oxon OX14 4RN

Simultaneously published in the USA and Canada
by Taylor & Francis Inc.
270 Madison Ave, New York, NY 10016

Routledge is an imprint of the Taylor & Francis Group

Typeset in Times New Roman by
Florence Production Ltd, Stoodleigh, Devon
Printed and bound in Great Britain by
The Cromwell Press, Trowbridge, Wiltshire

British Library Cataloguing in Publication Data
A catalogue record for this book is available from the British Library

Library of Congress Cataloging in Publication Data
Corey, Kenneth E.
Urban and regional technology planning: planning practice in the global knowledge economy / Kenneth E. Corey and Mark I. Wilson
p. cm. – (The networked cities series)
Includes bibliographical references and index.
1. Information technology – Economic aspects. 2. Knowledge management – Government policy. 3. Infrastructure (Economics)
I. Wilson, Mark I. II. Title. III. Series.
HC79.I55C675 2006 2005028735
303.48′33–dc22

ISBN10: 0–415–70140–6 (hbk)
ISBN10: 0–415–70141–4 (pbk)
ISBN10: 0–203–79943–7 (ebk)

ISBN13: 978–0–415–70140–2 (hbk)
ISBN13: 978–0–415–70141–4 (pbk)
ISBN13: 978–0–203–79943–7 (ebk)

Urban and Regional Technology Planning

Urban and Regional Technology Planning focuses on the practice of relational planning and the stimulation of local city-regional scale development planning in the context of the global knowledge economy and network society. Kenneth Corey and Mark Wilson explore the dynamics of technology-induced change that is taking place across many cities and regions as information society evolves. The book reveals to the reader the new local and regional planned development opportunities and potential of the global knowledge economy, laying special emphasis on the practicalities of the ALERT Model constructed by Corey and Wilson. This model suggests that local community stakeholders must raise their consciousness about the competitive position of the planning region in the global knowledge economy before local planners can engage in effective development planning today. Corey and Wilson focus their model on digital development and propose ways that representatives of the region can strategize for future intelligent development. Local and regional actor activation, empowerment and creativity are central to successful plan implementation under the ALERT Model.

Kenneth E. Corey is Professor in the Department of Geography and the Urban and Regional Planning Program of Michigan State University.

Mark I. Wilson is Associate Professor in the Urban and Regional Planning Program of the School of Planning, Design and Construction and the Department of Geography at Michigan State University, and also serves as a research economist with the Institute for Public Policy and Social Research at MSU.

The Networked Cities Series

Series Editors:
Richard E. Hanley
New York City College of Technology, City University of New York, US
Steve Graham
Department of Geography, Durham University, UK
Simon Marvin
SURF, Salford University, UK

From the earliest times, people settling in cities devised clever ways of moving things: the materials they needed to build shelters, the water and food they needed to survive, the tools they needed for their work, the armaments they needed for their protection – and ultimately, themselves. Twenty-first century urbanites are still moving things about, but now they employ networks to facilitate that movement – and the things they now move include electricity, capital, sounds, and images.

The Networked Cities Series has as its focus these local, global, physical, and virtual urban networks of movement. It is designed to offer scholars, practitioners, and decision-makers studies on the ways cities, technologies, and multiple forms of urban movement intersect and create the contemporary urban environment.

Mobile Technologies of the City
Edited by Mimi Sheller and John Urry

The Network Society
A New Context for Planning
Edited by Louis Albrechts and Seymour Mandelbaum

Moving People, Goods and Information in the 21st Century
The Cutting-Edge Infrastructures of Networked Cities
Edited by Richard E. Hanley

Digital Infrastructures
Enabling Civil and Environmental Systems Through Information Technology
Edited by Rae Zimmerman and Thomas Horan

Sustaining Urban Networks
The Social Diffusion of Large Technical Systems
Edited by Olivier Coutard, Richard E. Hanley, and Rae Zimmerman

In memory
Jean Gottmann   Gill-Chin Lim

For Marie, Jeff, Jen and Paul
For Lisa, Andrew and Kate
To our fathers, Ken and Ian

CONTENTS

Figures

PREFACE

The journey that brought us to this point – that of being able to share with others the lessons we have learned – began with our interest in the ways that information and communication technologies (ICT) were affecting people and places. Over twenty years of research and practice in ICT planning have shown us the importance of anticipating technological change, and harnessing these forces for community improvement. Information and communication technology is not the core issue in the book; technology has been a transformative agent in urban development, and urban and regional planning for centuries. The phenomenon of technological change is not new; what is important now is to shape and control ICT and capture the benefits it can offer our cities and regions.

Working with communities in the US state of Michigan and undertaking field research and collaboration with colleagues worldwide emphasizes for us the importance of creativity and action at a time of great economic change. The paradigms of industrial development that sustained growth in the twentieth century no longer apply in an increasingly integrated world economy. Cities and regions are affected by global economic and social changes, and need to be mindful of how their futures will be affected by these phenomena. Awareness of technological change in a global knowledge economy is not a simple choice, but a necessity if planners are to sustain healthy communities that offer opportunity, good standards of living and social justice.

The guiding principles underpinning the orientation of the book are prescriptive. It is not primarily a study or an analysis. In order to be effective in ensuring that this prescriptive guidance becomes imprinted in planning practice, fundamental change in mindset is essential. The intent of the book is to stimulate changed planning behavior at local and regional levels such that it is congruent with the new complexities and realities of the global knowledge economy and network society. As such, our primary goal is the creation and dissemination of actionable knowledge; information planners and citizen planners can use it to chart their own path through local development and technological change.

Kenneth E. Corey
Mark I. Wilson
East Lansing, June 2005

Acknowledgments

Many individuals and organizations assisted in the creation of this book. Much of the research was conducted as part of the Knowledge Economy Research Team (KERT) of Michigan State University's Community and Economic Development Program (CEDP). We thank CEDP Director, Dr Rex LaMore, for his extensive support and advice, and value our collaboration with John Melcher, Faron Supanich-Goldener and the members of the KERT team. The project benefited greatly from the support of the US EDA University Centers program 2000–05.

A number of MSU graduate students served as research assistants for the project, offering insights and expertise for data collection and mapping. Our thanks to Eric Frederick, Nicholas Helmholdt, Kyle Wilkes, Karan Singh and James Brueckman.

Data collection and research on ICT use was supported by a grant from the Institute for Public Policy and Social Research at Michigan State University, with invaluable assistance from the staff of the State of the State Survey.

Development of the book benefited from interaction with community members, planners and citizen planners in Michigan, who allowed us to test and refine concepts with those who will implement technology planning in a knowledge economy.

Our research and practice was supported intellectually through two valuable communities. First, E*Space: The Electronic Space Project is an international network of scholars and policy-makers committed to understanding the role of ICT in shaping space. We convened E*Space in 1994 and since then have met annually with colleagues to explore, understand and evaluate the global state of electronic space. The second community essential to our work is the International Geographical Union's Commission on the Geography of Information Society, chaired by Professor Aharon Kellerman, with Vice-Chair Henry Bakis and Executive Secretary Maria Paradiso.

Finally, our thanks to our families who, with grace and understanding, endured our preoccupation.

General Introduction

> If you give a man a fish, you will feed him once.
> If you take a man fishing, you will feed him for a week.
> If you teach a man to fish, he will never be hungry.
>
> (Lao Tse)

This book is written for the planner who wants to mobilize the region and its localities to engage the new development opportunities of the global knowledge economy and network society. The term "planner" is used broadly, capturing the ideas and work of the professional planner, the citizen or volunteer planner, and those with a commitment to the health and well-being of people and places. Ideally, the planner for whom this book is targeted is one who is willing and able to practice planning and invent the new practices that are required to engage successfully the development potential of the new economy and new society.

Urban and Regional Technology Planning is about the new practice of planning. It seeks to stimulate regional and urban planners to practice planning that is more likely to be effective and successful in the context of the global knowledge economy and network society. The book is not necessarily a formula. It is not a magic bullet. It is not a guidebook or a manual. Rather, the book intends to be a stimulus, prompted by the need for planners and citizens to learn how to create better places in a time of change.

We say that we want you to teach yourself, with your stakeholders and clients, how to be successful fishers within the context of the special and particular circumstances of your own local waters. Thereby, we want you to develop your own new approach to planning for the local engagement with the global knowledge economy. In order to assist you in this self-development and invention, our intention is to share with the reader some of the lessons that we have acquired on the journey of exploration we have taken. The goal of our journey has been to derive knowledge for use in enhancing the competitiveness of regions and localities by means of planning. From these examples and lessons, you may be prepared to better plan regionally and locally for the global knowledge economy and network society.

The book consequently is not a study. Rather, it is an example of how planners and their constituents may go about fashioning their own locally tailored approach designed to derive wealth and job growth from the global knowledge economy. From this example, it is intended that planners might be stimulated, indeed be inspired, to construct their own approach toward the global knowledge economy. Then planners would be positioned better to practice and perfect their relational approach for their region such that the new relational planning behavior, over time, becomes embedded in the routine planning practice of local planners and their stakeholders. By adopting this stance, it is recognized that, in practical terms, places are nearly unique, and each might forge its own particular path of development within the constantly changing context of the global knowledge economy and network society. Thus, it should be recognized that there is no one formula or recipe for doing regional and local relational planning. In the context of the particularities of relatively unique places, it should be acknowledged that local planning must be a do-it-yourself and experimental activity.

The bulk of the ideas and lessons discussed in the book has come from the work that has been ongoing over the last five or six years in seeking to model ways that the substate planning regions of the US state of Michigan might effectively relate to the global knowledge economy and network society. From the three major technology-economic regions of the global economy, parallel cases have been identified and used to inform the development and evolution of the regional relational-planning approaches and models introduced here.

Given the "relative uniqueness" or particularity of places, there is no attempt here to be comprehensive and exhaustive, or even representative, in coverage. It is simply too early in the evolution of regional and local relational-planning practice for that. The intent here is to suggest ways to operationalize regional relational planning, and thereby contribute to the improvement of contemporary practice of regional and urban planning. While it is intended to root the discussion in theory and proven concepts, the goal also is to keep the discussion simple and straightforward. The modest intent is to share experience and experiment. From such sharing and incremental local experimentation, reasonable and operational planning approaches and content successes are possible, indeed even probable.

One of the most important discoveries that we uncovered on our journey of exploration was relational theory and its application, relational planning. Given that the effective regional relational planner needs to provide a close working interdependent leadership with the region's principal stakeholders, many of the references cited include popular publications, so that a wide range of stakeholders might inform themselves and be able to work with the local planner. Thus, both planners and stakeholders are learning to operationalize and tailor local relational planning together.

The book is divided into five parts. The first part on the knowledge economy aims to identify the economic and social forces shaping life in cities and regions, and to provide an understanding of the nature of the knowledge economy, and its implications for cities and regions. Our primary interest is in developing awareness of the role that planning must play in the knowledge economy. In Part II, emphasis is on the role of fundamental planning concepts and the need for a new mindset to tackle the issues of urban and regional development in a knowledge economy. One way of achieving a new mindset is the application of spatial relational-planning concepts to contemporary planning. Part III addresses the role of policy, and draws on examples of local planning from the globe's three principal technology-economic regions that respond to the needs of the knowledge economy. Part IV offers the core of the book, the presentation of the ALERT Model, which applies relational-planning concepts to regional and local planning practice in the context of the global knowledge economy and network society. Finally, in order to keep the main narrative of the book uncluttered, the reader is referred to various sections of Part V which includes support materials, terms, definitions and tables to complement the body of the book's discussion.

After reading the book, it is our hope that planners and other stakeholders in cities and regions will think anew about how they approach planning, and how they might find the best path for their communities given the nature of the knowledge economy and network society.

PART I
Change

Introduction

One defining element of any society is the nature of the economy and the work performed by people each day. Of growing importance is the preeminence of knowledge and information that has become central to the many economic, social, and political actions that shape our lives. While knowledge and information always have been important and played a role in economic growth and development, the current evolution of advanced economies places a premium on the generation and management of knowledge. The shift from the physicality of agriculture and manufacturing to the intangibility of information carries with it significant changes in the occupations and industries of most economies and societies. Changes in what we make and the skills valued in the marketplace carry implications for individual well-being, as well as for the welfare of the places created by the industrial age that must now adapt to a new form of production. In this context of change, planners must act to harness the advantages of the knowledge economy, while trying to mitigate the disadvantages for their communities.

Economies often are identified by their dominant activity, such as agriculture, manufacturing or services. As the importance of services, information, and knowledge has grown, many advanced economies now see themselves by these terms. The information age, or knowledge economy, is a common term applied by planners and analysts to the current state of the economy. Simply stated, knowledge has emerged as a significant input to production, in addition to labor and capital, and increasingly is seen as the driving force for local, national, and global economic development.

While the form of the current economy today in many countries is different from the past, the process is not new. The evolution of economies from agriculture to manufacturing to services have been observed and acknowledged since the 1930s, with scholars first studying the industrialization process in the nineteenth century in Europe and North America, and later recognizing the rise of services from the 1950s. Fisher (1939) and Clark (1940) recognized a three-stage economy, comprising primary (agriculture), secondary (mining and manufacturing) and tertiary (services) activities. The early classification of economic activities led to an expectation of a linear development process through all stages, with a tertiary or service economy the ultimate goal. The process of economic evolution is far more complex than initially suggested, with some developing countries having dominant tertiary sectors that are closer to a starting point than an economic goal. In addition, as the nature of the economy changed, not only did the character of industry evolve, but also the nature of many occupations. The shift from agriculture to manufacturing, the industrialization process, is inextricably associated with the rise of urbanization. Andersen and Corley (2003) remind us, however, of the problem of relying too much on this classification system, and avoiding the assessment of services as unproductive or immaterial.

The breadth of tertiary services, which were a small part of the economy in the 1930s but far more significant later, led to further disaggregation of the category into tertiary, quaternary and quinary services. This division recognized the wide range of services emerging in the economy. Tertiary services refer to telecommunications and transportation, with quaternary services being provided by professionals, such as banking and legal services, and quinary services being provided by such organizations as medicine, education and government. This expanded categorization maintained the expected linear progression, with the most advanced economies expected to be dominant in quinary services. This form of classification is useful as it shows how economies evolve in a general way, although it serves us better as a descriptor than as a policy prescription.

During the 1960s, 1970s and 1980s, scholars and policy-makers focused attention on the nature of the service economy and its implications for growth and well-being. Major issues arising from this economic evolution include measuring services productivity (Baumol and Bowen 1966), conceptualizing services as an intangible economic function, the location of services, the deskilling process (Braverman 1974), and the problem of workers displaced as manufacturing jobs were lost and replaced by service occupations (Harrison and Bluestone 1988; Reich 1991). The shift to a service economy thirty years ago was accompanied by a number of issues we still confront, such as the nature and trajectory of the economy, and the ability of communities to regain past high living standards delivered by manufacturing.

Among the past issues that remain relevant is the impact on employment of services, and elements such as worker displacement as manufacturing jobs were lost and employment growth concentrated in service occupations. The loss of manufacturing jobs continues in many advanced economies, with service growth divided between low-paid unskilled work and well-paid highly skilled occupations. The middle, career-path jobs supplied by manufacturing have been lost and there is little in the current economy to take their place. Policy-makers debated the nature of service work, with debate over the intangible service being seen as unproductive and irrelevant compared to the solidity of manufacturing industry. There was opposition to services by some, not because of their polarization of wages but because of a sense that services were unproductive and unable to support an economy. While services are seen now as being legitimate bases for economic growth and development, we still need to confront the polarization effects in the labor market.

Another element of the service economy that has been reassessed during the past thirty years is its mobility. As manufacturing employment was lost to Asia, Mexico and Latin America, many felt that service work would remain in place because the nature of services did not lend itself to mobility. In one sense, this is true and remains valid. Personal services, such as restaurants, hairdressers, theaters and musical performances, and medical facilities need to be taken advantage of in the location where the services are offered. Increasingly, however, many services are being relocated away from high-wage areas in North America, Europe and Japan to low-cost locations in Asia and Mexico. Among the first examples of this trend was the move offshore of back-office functions to the Caribbean, Ireland and the Philippines (Wilson 1995; Wilson 1998).

Basic paperwork, information processing, data entry, software development and animation left, and continue to leave, traditional production centers for low-cost locations offshore. A range of intermediate services originally produced in the United States, Canada, Britain, Singapore and Hong Kong now are found in Jamaica, India, Ireland, China and Barbados. The simplification and mass production of back-office services, combined with the spatial flexibility afforded by communications, changed the structure and location pattern of the global service economy. Despite the offshore moves of basic services, many analysts felt comfort in the fact that advanced, well-paid, service jobs would remain in the markets they served.

During the 1980s, as more advanced service functions relocated offshore, scholars analyzed a far more nuanced service economy. Services were not one massive sector, but a wide range of jobs from very low skilled to highly advanced occupations. The treatment of services as one industry remains with us in many ways; the very way we treat and present economic data ignores many of the important distinctions of the sector. Scholars analyzing services and globalization, such as Daniels (1985), Feketekuty (1987), Harvey (1989), Warf (1989), Hepworth (1990), and Sayer and Walker (1992) noted the growing ability of services to exhibit the mobility experienced earlier by manufacturing, and the changing organization and control of work in a post-Fordist economy.

The primary motivation for offshore relocation or outsourcing of services is access to low-cost labor. Labor cost for office work is not wages alone, but a variety of other factors that influence worker productivity and employer expenses. In addition to wages, labor costs are affected by productivity and worker turnover, worker availability, benefit packages, occupational safety and health considerations, and cultural and social influences. Labor-intensive office work in the US faces wage and cost pressures from declining numbers of young workers and increasing insurance and benefit costs that are not evident in many offshore locations. The resulting high wages and expectations of future wage growth has made low-cost labor the central issue for many service functions. The appeal of low-cost labor in distant locations was one of the factors that encouraged the development of ICT infrastructure.

The growing mobility of services was aided greatly by the development of information and communication technologies to routinize and free many forms of service work. While it remains true that one has to be present to get a haircut, ICT meant that financial advice and medical information could be obtained remotely. The crucial element in understanding the mobility of services was distilled by Nicholas Negroponte (1995) as the difference between atoms and bits. While atoms – meaning manufactured goods – were tangible and needed to be physically assembled and moved around the world, the nature of information in many services was intangible and able to be moved electronically, by bits. For many products, ICT freed service production and consumption, and allowed the growing spatial fragmentation of service employment.

In an early *Economist* article (Cairncross 1995) and later book, Cairncross (1997) captured the impact of ICT as *The Death of Distance*. ICT was able to conquer distance for services as container shipping and aviation had revolutionized the supply chains of manufacturing. At this stage, it is important to note the distinction between distance and place, for while distance may well be compressed or annihilated by electronic media, the qualities of place remain more important than ever. Harvey (1989: 124) reminds us that "The problem of space is not eliminated but intensified by the crumbling of spatial barriers." The reason that conquering distance is so important is to gain access to places with attributes of low costs, technical expertise, or desirable markets. Places carry social, cultural, economic and political characteristics of amazing heterogeneity, rewarding the effort needed to establish electronic and commercial linkages. If places were not very different, then the need to conquer distance would be minimized. Of course, electronic access may well be the means to diminish difference, to be the force that thrives on place-specific characteristics at the same time that it serves to minimize them.

In the past twenty years, the five-stage economy and simple manufacturing/services distinction has lost popularity as an organizing and policy framework, as attention focuses on advanced services and later the knowledge economy. Among the changes being experienced as part of structural change are the rise of research and development functions; the growing importance of education and training to meet demands for workers who are able to contribute value-added in a knowledge economy; and the polarization of population into those who are able to obtain education and benefit from a knowledge economy, and those who are isolated and exempt from its benefits. The evolution from an agricultural to an industrial society required the adoption of new skills, and also a spatial reorganization of production, with cities being far more effective means for the organization of production. As we move from an industrial to a knowledge-based economy, we need to prepare for the social and spatial changes that might accompany this powerful economic realignment. The move from farm to factory to office, and from tangible to intangible production, characterizes the changing condition of many economies during the past decades.

Before designing policy and implementing action, it is essential to understand the processes that shape any community. Awareness of recent trends, especially those associated with the economy and technology, is an essential first step. This part of the book will identify several trends that planners need to understand and utilize in their professional work. Among the trends associated with contemporary planning are (1) the nature of the knowledge economy; (2) the role of information technology; (3) the spatial organization of the global economy; and (4) issues for the knowledge economy.

The Knowledge Economy

Changes in the way production is organized carries with it major social, economic and political change for work, workers and the places where we live and work. Economists, political scientists, planners and sociologists long have recognized the importance of linking the nature of work to the condition of our lives and our daily well-being. The process of structural change, by which an economy moves through phases of different industrial and occupational emphasis, demands our attention if we are to understand and then shape outcomes. Historically, we have chosen to associate periods and stages of development in terms of the prevailing economic function, such as agrarian, industrial and post-industrial eras and societies. Information and knowledge always have been important to production, but since the Industrial Revolution its importance has intensified and become dominant. Mokyr (2002) notes how science and technology fueled the Industrial Revolution and led to sustained growth in many of the world's economies. This process continues, with science and technology gaining increasing importance as a driver of the economy.

The preeminence of information and knowledge as social and economic factors was recognized by historians and economists a century ago, and received increasing attention during the latter half of the twentieth century when the shift from manufacturing to services employment prompted detailed analysis. Structural change in advanced economies during the 1970s was recognized initially as the transition to a service economy in which the role of knowledge was implicit. Toffler (1970) and Bell (1974), among others, forecast major social change associated with the transition to a services economy. Bell presented a postindustrial society where science and technology rose to a central position to fuel a services economy based on knowledge. Bell correctly predicted the rise of professional and technical classes as a result of the need for an educated workforce, yet the expectation that knowledge work would provide a better life for workers is tempered by the experience of deskilling and outsourcing during the past decades when skilled workers did benefit, but those with less education gained little from the knowledge economy.

At the same time, analysts, such as Gershuny (1978), raised questions about the negative implications of the knowledge economy, such as its power to alienate and divide groups within society.

During the 1980s, the growing importance of technology and advanced processes were recognized and sought after as engines of economic growth. In part, the focus was on manufacturing and ways to innovate in the design and production of industrial and consumer commodities. Ann Markusen, Peter Hall and Amy Glasmeier (1986) wrote *High Tech America* at this time and noted the importance of this industry and the breadth of its impact. Of significance were their findings of the large scale of job creation in high-tech manufacturing and the ability of this sector to develop in both new and old industrial cities and regions. By recognizing the ability of many areas to offer appropriate conditions for high-tech industry, the authors also illustrated the mobility of these economic activities, a prescient recognition of the globalization of production to come during the 1980s and 1990s.

Recently, attention on high-tech manufacturing as a driver of the economy has given way to a focus on the knowledge component alone. The fragmentation of production and the use of ICT to link global production centers allows the knowledge component of production to be isolated and emphasized. The importance of knowledge to the economy should not come as a surprise. Economics long has recognized the importance of human capital to economic success, and planners and community development specialists have tried to upgrade the skills in their areas to support better jobs and attract employers. The loss of middle-income, career-path jobs evident since the 1970s is part of this trend. In the past, jobs lost could be replaced with jobs of similar skill levels. The knowledge economy's sobering quality is the premium it places on education, and the limited opportunities it offers for those with limited education or skills.

Emerging from analysis of the knowledge economy has been recognition by some of the role of creativity as the force behind knowledge. Landry (2000) directed planners to think about the cultural assets of their communities as an economic resource, able to spur innovation and development. Richard Florida (2002b; 2005b) raised the profile of urban creativity further with *The Rise of the Creative Class* and more recently *The Flight of the Creative Class*, with its emphasis on the globalization of knowledge jobs away from the United States. Florida rightly draws attention to the power of cities to be crucibles of ideas, innovation, and creativity, and the power of diversity and tolerance as attractors for creative people (2005a). At the same time, it is essential to remember that at the core of the knowledge economy is education, and the need for communities to address this crucial factor first before seeking ways to be cool, hip or Bohemian. The value of the recent attention paid to creativity is that it serves as a reminder of the power of innovation and knowledge, but it can also be distracting, with energy expended on creativity as an end rather than a means.

Structural Change

The scale and scope of change associated with the knowledge economy can be illustrated by changing employment patterns for both industries and occupations, and rising education requirements for employment. In many countries in North America, Europe and Asia, development has brought changes in sectoral significance, with employment declining in agriculture and manufacturing, and increasing in services (International Labour Organization 2004; Organisation for Economic Co-operation and Development 2003).

Industry

The industrial structure of an economy shows the types of goods and services produced, such as manufacturing or construction. As the advanced economies of North America, Europe and Asia have developed over the past decades, employment needs have shifted away from agricultural and manufacturing work toward service industries such as finance, wholesale/retail trade, health, education and transportation. The economic structure of 25 countries of the Organisation for Economic Co-operation and Development (OECD 2003) is presented in Figure 1. Data show that in most of the countries listed, agricultural employment is less than 5 percent of total employment, with manufacturing often less than a quarter. Within the OECD, a number of countries report low employment for agriculture, such as the United Kingdom (1.3 percent), the United States (1.9 percent) and Canada (2.8 percent), with service employment exceeding almost 85 percent in the United States and over 75 percent in the United Kingdom, Sweden and Luxembourg. The economies of southern and eastern Europe, Mexico and Turkey show larger roles for agriculture and manufacturing, but with development these sectors will shrink as services take a larger role in the economy.

Due to the lack of consistent labor-market definitions, it is difficult to compare detailed employment levels across many countries, but for any location – city, region, state/province, country – it is possible to identify major industries and employment trends. For example, the experience of the United States over the past twenty years shows how industrial structure can change over time. Figure 2 illustrates the industrial structure of US employment for 1980 and 2000. Most significant during this period is the decline in manufacturing employment and the rise of services, such as business services (advertising, computer/data processing, etc.); personal services; and professional services (health, education, social and legal services). Within the broad services category, economies evolve to concentrate on knowledge activities associated with production (education, research and development), and basic services such as retail trade and transportation.

For planners, awareness of industrial employment trends in their city or region will show sectors of promise or concern. While at the national level trends to services are consistently found, experience may be different locally. Cities and regions may depend on agriculture or manufacturing as their driving

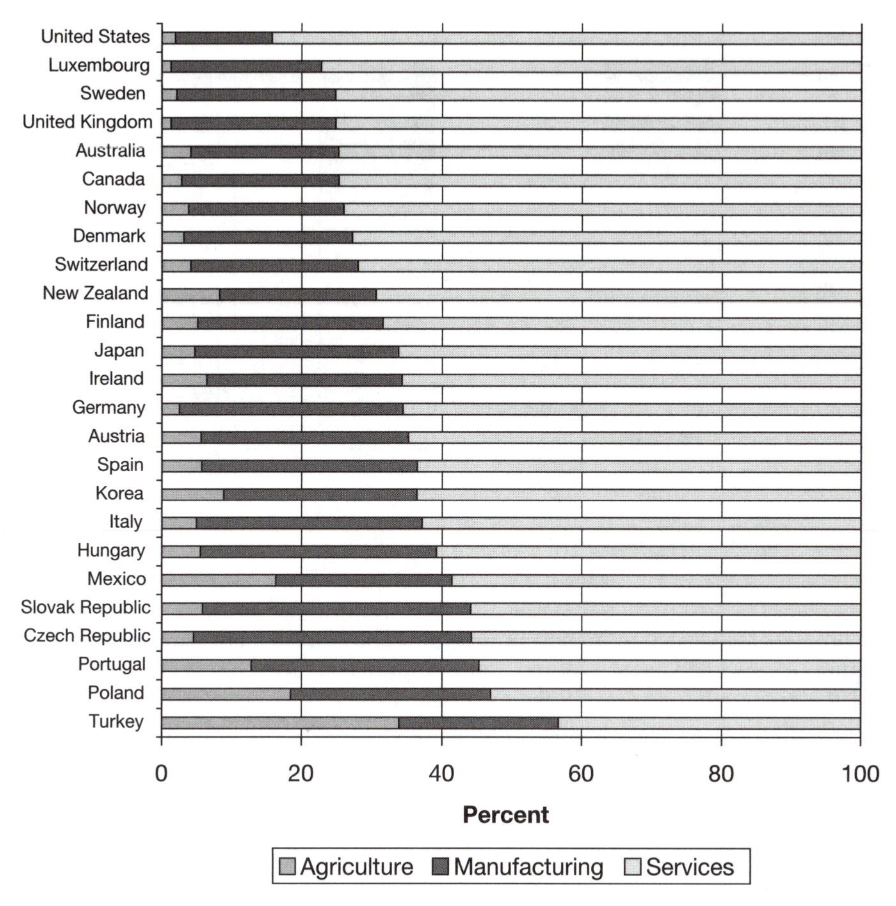

Figure 1 Employment by Sector and Country (2003)

Source: Organisation for Economic Co-operation and Development: Science, Technology and Industry Scoreboard 2003 – Towards a Knowledge-based Economy.

economic force, and this specialization needs to be understood through detailed analysis. Employment data by location is an essential input into the planning process, although not all locations have accurate data available. Planners confront problems of outdated information or lack of disaggregation by industry or activity for their region.

Occupation

In contrast to industry structure, which shows what an economy produces, occupational structure shows the skills needed by an economy for production. In fact, the shift to a knowledge economy places emphasis on education and the skills of the individual, making occupation of increasing interest to analysts. Just as the types of goods and services being produced has changed, so have

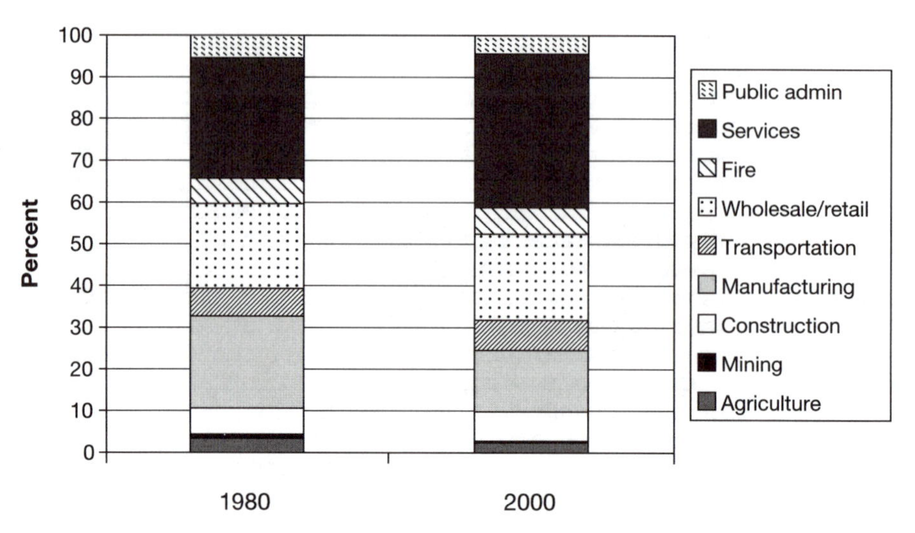

Figure 2 *Employment by Industry, USA (1980 and 2000)*
Source: United States Census, Statistical Abstract of the United States (2001).

the types of skills needed in their production. The shift to a knowledge economy requires more knowledge workers, and employment to support these workers and the knowledge infrastructure (National Research Council 1999).

Paralleling the shift to advanced services in terms of industry has been a shift to occupations with higher management and skill levels. This trend is exemplified by the experience of the United States, which shows employment growth for managerial and professional occupations, including jobs such as managers, engineers, teachers, lawyers, scientists and physicians. Figure 3 shows the occupational structure of the US economy for 1983 and 2000. The strongest growth occurs for managerial and professional occupations, which, by 2000, account for almost one-third of the occupations in the US economy. Losses occur for farmers, now only 2.5 percent of occupations, and for production workers and operators/fabricators who usually are tied to manufacturing. A similar trend is evident for Europe, with a 2004 study by Eurostat finding:

> From 1995 the number of workers increased by 32% among technicians and associate professionals and by 24% among service workers, but only by 1% among craft and related workers. The only occupational group showing a decrease in workforce was skilled agricultural workers (a decrease of 11%).
>
> (Eurostat 2004: 21)

Comparisons of occupations across countries is complicated by lack of consistent definitions and collection, but an example of how occupational structure varies is illustrated in Figure 4, showing employment by occupation

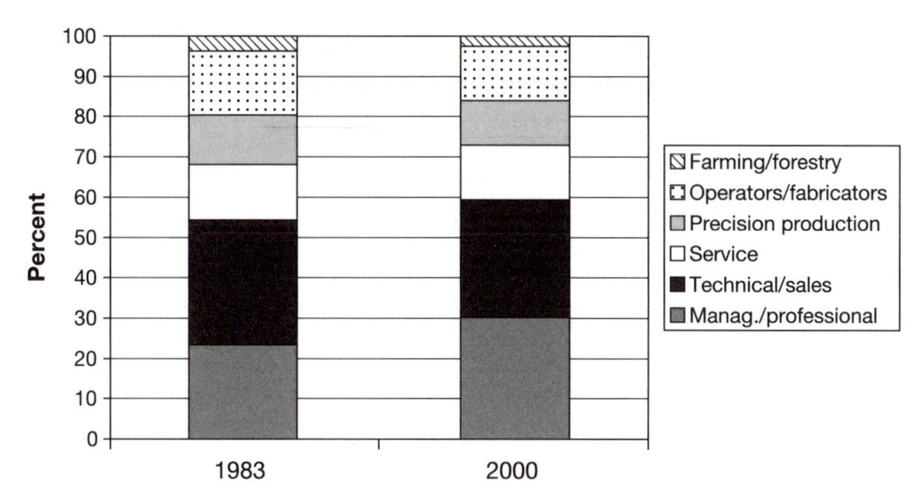

Figure 3 *Employment by Occupation, USA (1983 and 2000)*
Source: United States Census, Statistical Abstract of the United States (2001).

in 2003. For the six advanced economies analyzed, management, professional and technical occupations account for more than one-third of employment, with a similar level for manufacturing oriented production occupations.

The occupations that are projected to expand in the coming years also illustrate shifting occupational structures and the growing importance of an educated workforce. For the United States, the Bureau of Labor Statistics has identified the fastest growing occupations for the period 1998–2008. Of the 30 fastest growing occupations, 20 require an associate's, bachelor's or graduate degree, with the remaining jobs needing on the job training. The fastest growing occupations include such specialties as computer engineers, systems analysts, paralegals, medical assistants and data processing. Of the 30 occupations with the greatest job growth, 9 require some college education as a minimum. The largest growth in absolute numbers is anticipated from a wide range of occupations such as systems analysts, retail sales, cashiers, managers, truck drivers and clerks. In addition to underscoring the importance of education for many of the fastest growing occupations, data also remind us that there will be considerable growth in low-skill occupations as well. The implication of this is a divided society with highly paid, educated knowledge workers contrasting low-paid, basic service workers.

Education

As the economy rewards skilled and better educated workers, the educational attainment of the workforce has increased steadily over the past decades. The ability for high-school graduates to find well-paid career paths in manufacturing was a hallmark of life in most advanced economies during the latter half of the twentieth century, yet this outcome had become rare as the economies

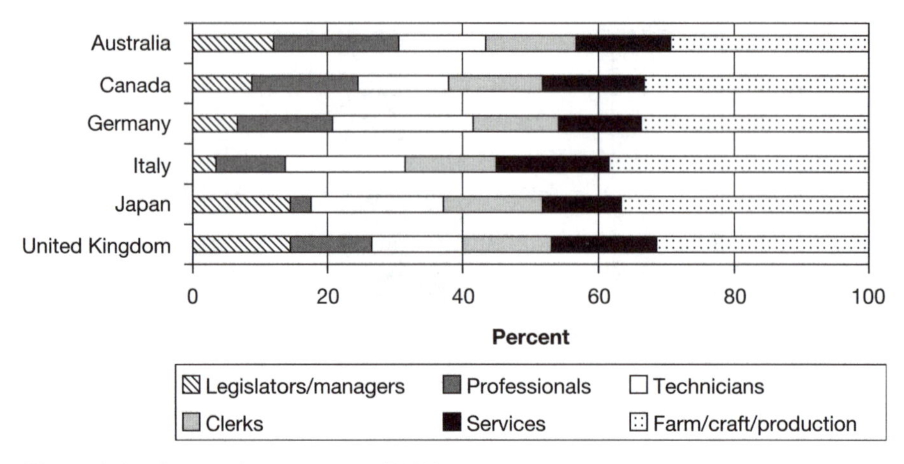

Figure 4 *Employment by occupation (2003)*

Source: International Labour Organization: Labour Statistics Database (2004).

of the 1980s and later demanded more skills. Across the OECD countries, tertiary education was claimed by 28.2 percent of the workforce, ranging from a low of 9.9 percent in Portugal to a high of 41.9 percent in Canada (OECD 2003). Employment and income growth across the OECD countries also is higher for those with tertiary education. Of the residents of the European Union in 2002 (EU-15), an average of almost a quarter has tertiary education qualifications, and in many cases this exceeds one-third of those aged 25–34 years (Eurostat 2003). In the United States, of 110 million workers in 2000, almost one-third (34 million) had college degrees, with more than half of the workforce (65 million) having attended or completed college. Of the growing occupations, such as managerial/professional, college graduates accounted for almost two-thirds (64 percent) of the occupation group, while declining occupations such as precision production and operators/fabricators have more than two-thirds of their workers with a high-school diploma or some high-school education.

The Role of Information and Communication Technology

Much of the character of the information age owes its existence to technologies such as computers, telecommunications networks, electronic media and the Internet. Computers provide the power to generate, manipulate, manage and index information. Telecommunications allows the movement of voice and data across space, while the information economy is cause and effect of the development and evolution of both computers and telecommunications. The importance of information is tied closely to the development of technologies to gather and manage this resource. Information technologies, initially computers and later the Internet, were developed to improve information processing, offering new ways to collect and manage information. As information was mined for its value, emphasis started to shift in its economic valuation, with core information growing in value relative to the traditional products of the manufacturing era. Stephen Saxby (1990) underscores the significance of technology and information, noting that the marriage of the two has

> provided a stimulus that can be seen not just in terms of new products, services, wealth and success for the richer nations, but in the far-reaching implications for business methods, design and manufacturing techniques and the way individuals interact, travel, entertain themselves, obtain information and communicate.
>
> (Saxby 1990: 3)

The significance of this change is examined in detail by Manuel Castells (1996) who opens *The Rise of the Network Society* by recognizing a "technological revolution" that is "reshaping, at accelerated pace, the material basis of society." The implications of this revolution range widely, affecting many elements of our daily lives, such as the work we do, the sources of information we receive, the spatial structure of production, and the growing division between those who benefit from a knowledge economy, and those who cannot find a role or who are excluded.

The intersection of the knowledge economy and ICT can be seen from many dimensions, such as its form as an economy (Negroponte 1995; Shapiro and Varian 1999; Evans and Wurster 2000), with new opportunities for the organization of production, the structure and operation of organizations, and new products and services enabled by e-commerce. At the personal level, the relationship between individuals and information and communication technologies offers new forms of social interaction and virtual communities (Rheingold 1993; Turkle 1995). Changes in production, economic organization and social interaction also require institutional change in the legal and political context as noted by Stefik (1999) and Biegel (2001).

The evolution of ICT often is seen as two distinct spheres that are mistakenly considered as separate, the engineering-oriented application of hardware and the social impact of new technology. As ICT grows in significance, it is clear that the commercial application and use of the technologies represents many social, economic and political spheres that constantly overlap. The social context of technology and its impact long has been recognized, although often lessons learned from the past are not applied. Experience with ICT policy over the past decade shows a preoccupation with the technology and far less interest in its application. Many communities focus more on the means – i.e. the ICT – than the ends – i.e. the services they can use and leverage to improve daily life. To provide a framework to investigate ICT and its planning, it is valuable to recognize three distinct, but interrelated, facets of information and communication technology: technical possibilities, economic viability and social acceptability. The technical and economic aspects tend to receive most attention, yet it is the social context that may well determine the success of the ICT choices made for and by our communities.

The first dimension of ICT is the technical possibilities it offers. Primarily, this is an engineering concern, seeking new electronic and scientific advances to improve performance, reduce costs and produce new applications. ICT advances are spurred by basic research in science and engineering, and the economic advantages of developing and taking new ideas out of the laboratory. The technical possibilities offered by information technologies tend to be produced by the world's research and development centers, which tend to be located in North America, Western Europe and several Asian nations (Japan, Taiwan, South Korea and Singapore).

While many policy-makers, planners and the technology industries are focused on technical issues, these technical possibilities alone should not be our sole concern. Computing power and telecommunications may be available, but that is only part of the formula; if the cost is too great, use and impact will be limited. The declining cost of international telephone calls exemplifies the demand boom associated with drastic cost reductions by telecommunications providers. As computer production is moved to low-cost assembly locations, the cost of computing power also declines. The combination of technical advance and production reorganization means that the cost of ICT use is dropping, although many individuals may still be prevented by cost from accessing ICT.

The social or behavioral dimension of ICT is perhaps the most powerful force for adoption and use, yet it also tends to receive less attention than technical and economic issues, even though all often intersect. The social element of ICT use has many facets. One important element is the relationship between individuals and technology, and the ways that different individuals, groups, cultures and societies perceive, apply and use information and communications technologies. Another dimension is the relationship between technology and the state, and the ways in which government affects ICT use through industry – e.g. subsidy, regulation, standards – and policy – e.g. taxation of Internet commerce, cyberlaws, etc. A related element is state and society relationships, which introduce technology issues of access, ownership and control, privacy and security.

Access to ICT

Internet access is moving from a convenience to a necessity for functioning in advanced economies. As the range of Internet services expands, those able to access material online will have a substantial information and financial advantage. Among the advantages to those with Internet access are: (1) information available online faster, or only available in electronic format; (2) reduced costs through online transactions, such as renewing permits, paying bills, or working from home; and (3) reduced costs of services, such as travel or use of online auctions. Access to the Internet, and therefore access to the services it provides, is affected by a range of factors, starting with the individual's interest in the Internet and willingness to purchase access or visit public sites such as schools and libraries.

Internet access usually comprises two costs, an Internet Service Provider (ISP) subscription and the cost of connecting to the ISP. ISP subscriptions are either an hourly rate or a set rate for unlimited access. With unlimited access, users have the freedom to consume bandwidth and time exploring and using the Web, while hourly users face an increasing cost as their time online increases. Countries with low-cost unlimited access will have a more sophisticated Internet user because the pricing structure allows greater familiarity with the Internet. Surveys across different countries will show different Internet access rates for various reasons, but the cost of access needs to be incorporated into comparative measurement. For example, Figure 5 shows the cost of Internet access for 40 hours, with the cost of telephone and ISP charges ranging from US$29.40 in South Korea to US$156.08 in the Czech Republic in US dollars adjusted for purchasing power parity. The European Union average was $66.34 and the OECD average was $64.20. Costs varied with development status, as both Turkey and Japan had Internet access costs of approximately US$49.

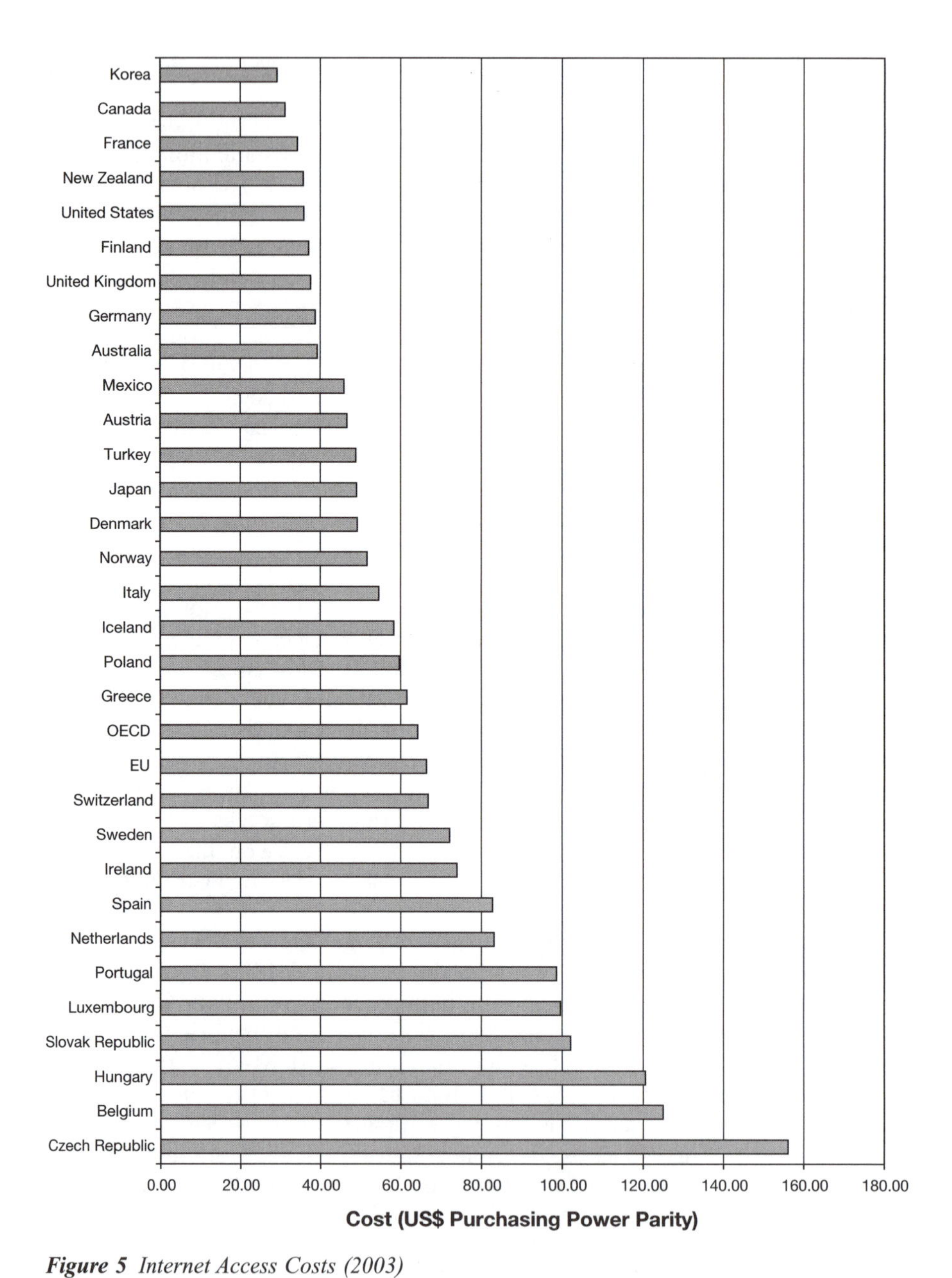

Figure 5 *Internet Access Costs (2003)*

Source: Organisation for Economic Co-operation and Development, Telecommunications Database 2003.

Equally important in many countries is the cost of the telephone call to access an ISP. Countries with telecommunications services that allow an unlimited length call for a local charge, such as Australia, Canada, or the United States, provide low-cost access to ISPs. Countries with metered local telephone charges, or people living in locations that require a long-distance call to reach an ISP, face far higher access costs.

Although commercial ISPs are readily available almost everywhere in advanced economies, they are still a rare commodity in many parts of the world, while telecommunications costs can be as significant an element of access as ISP service availability. Access to the Internet is a complex matter, as it is not solely an issue of the availability of ISPs, but their cost and the costs of accessing services. The factors that shape access go beyond the technical and also include government regulation, existing patterns of telecommunications charges, and the level of competition for both ISPs and telecommunications providers.

Information Technology Infrastructure

As the underlying technology of the knowledge economy, telecommunications and computers merit analysis as an enabling infrastructure. ICT often is portrayed as a universal set of technologies that can be found and used anywhere. Narrowly interpreted, it is possible to access the Internet and use computers anywhere on the planet through satellite communications. More broadly, however, ICT remains highly concentrated in advanced economies and metropolitan areas. The lower costs of technology and communications are improving access, but there still remain many places where even a telephone cannot be found. Access to ICT is one element in the preparation for the knowledge economy, and in this section we will address the scale and scope of global ICT infrastructure.

Worldwide in 2003, for more than 6 billion people there were 2.5 billion telephone subscriptions (both line and mobile) according to the ITU or the International Telecommunications Union (2005), or 41 subscriptions per 100 inhabitants of the planet. This is a rapid increase from just two years earlier when there were 2 billion subscribers worldwide, or 33 subscriptions per 100 inhabitants.

Data on telephone subscribers by country for 2003 are mapped in Figure 6. In a broad definition of Europe with 40 countries, the ITU estimates subscriptions average one per inhabitant, followed by Oceania (95.2 subscribers), the Americas (68.9 subscribers), Asia (29.1 subscribers) and Africa (8.7 subscribers). In advanced economies there is more than one subscription per person, such as Luxembourg (199.1 subscriptions per 100 inhabitants); Taiwan (173.2 subscriptions); Hong Kong (163.8 subscriptions); Sweden (162.4 subscriptions); and Norway (162.2 subscriptions). High subscription rates are evident also for

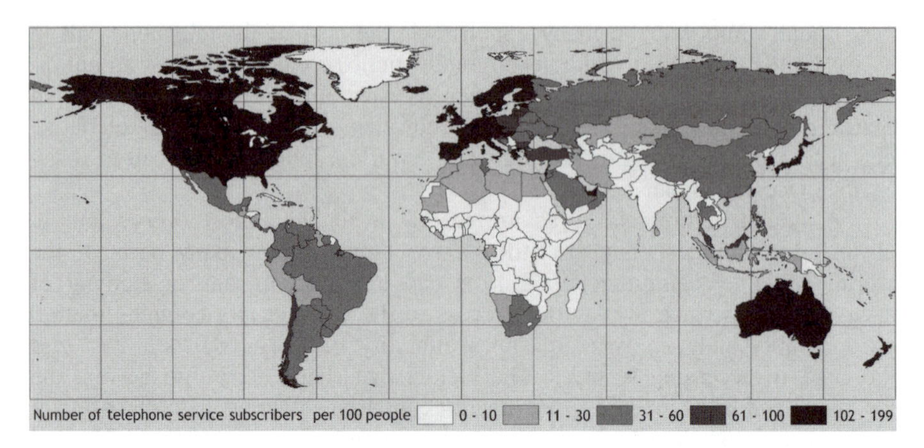

Figure 6 *Telephone Subscribers (2003)*

Source: International Telecommunication Union (2005), main telephone lines, subscribers per 100 people.

major economies such as Germany (144.3 subscribers), United Kingdom (143.1 subscribers), France (126.2 subscribers), the United States (117.0 subscribers), Japan (115.1 subscribers), and Canada (107.0). Contrasting the abundance of telephone service in developed economies, there remain many places with very low levels of telephony. Of note is Africa, which has 71 million subscriptions for over 800 million people, and almost half of these subscriptions are for South Africa and Egypt. Many countries have fewer than one subscription per 100 people, such as Niger (0.3 subscriptions), Chad (0.6 subscriptions), Myanmar (0.8 subscriptions), and Ethiopia (0.8 subscriptions). Telephony does vary considerably among developing economies, such as the contrast of India (7.1 subscriptions) and China (42.4 subscriptions).

Telephone service consists of wired lines and cellular subscriptions, presented in Figure 7 and Figure 8. In 2003, there were over 1.1 billion lines worldwide, averaging 18.7 lines per 100 inhabitants. Wired services ranged from Europe (41.2 lines), to Oceania (40.7 lines), the Americas (34.2 lines), Asia (13.4 lines) and Africa (3.0 lines). As the earlier technology, telephone lines are more common in countries achieving development earlier, such as Western Europe, the United States and Canada. The cost and logistics of wired telephony often created delays and poor service in developing countries, and the roll-out of the technology posed a significant barrier to ICT infrastructure development.

The development of cellular or mobile technologies offered new ways of delivering telephone services. In 2003 there were 1.4 billion cellular subscribers or 22.9 subscriptions per 100 inhabitants, accounting for more than half, 55 percent, of all telephone subscriptions. Cellular subscription rates among major nations include United Kingdom (91.2 subscriptions), Germany (78.5 subscriptions), France (69.6 subscriptions), Japan (67.9 subscriptions), United States

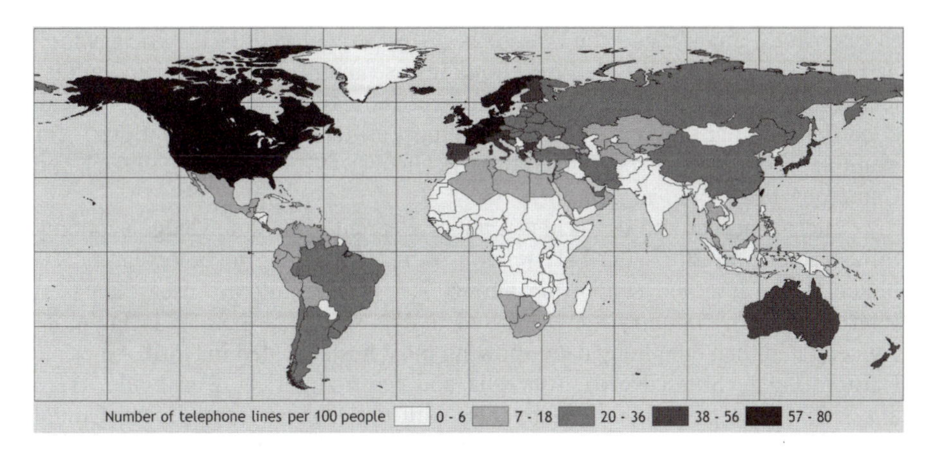

***Figure** 7 Telephone Lines (2003)*

Source: International Telecommunication Union (2005), main telephone lines, subscribers per 100 people.

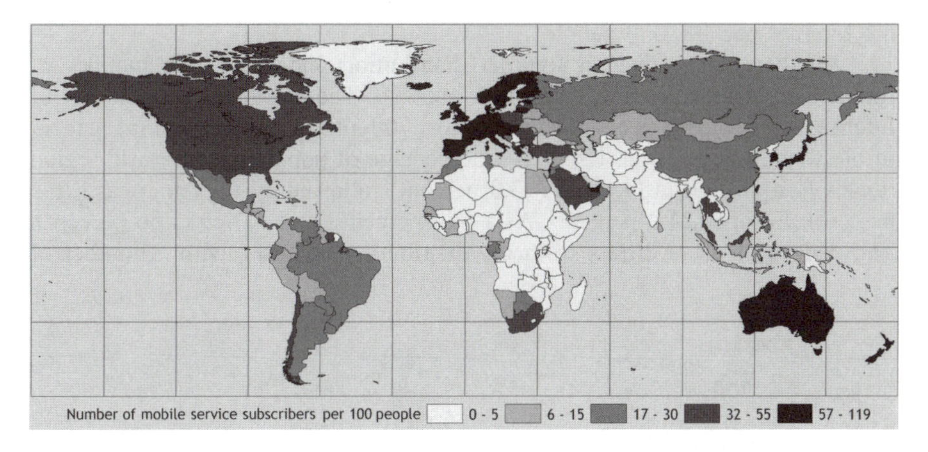

***Figure** 8 Cellular Subscriptions (2003)*

Source: International Telecommunication Union (2005), mobile cellular, subscribers per 100 people.

(54.6 subscriptions), Canada (41.9 subscriptions), China (21.5 subscriptions), and India (2.5 subscriptions).

Cellular services cost less to deploy due to the reduced infrastructure needed, especially the last mile of wiring to houses and businesses. Countries with well-developed wired infrastructure had less incentive to adopt the new technology, while developing countries that did not have a significant wired infrastructure adopted cellular technology to extend service. In all major world regions, cellular services exceed wired service use: Africa (67.5 percent cellular), Europe (58.9 percent cellular), Oceania (57.2 percent cellular), Asia

(54 percent cellular), and the Americas (50.3 percent cellular). Africa, which had the least developed wired telephone infrastructure, benefited from cellular technology and today relies significantly on cellular service. In many African countries, cellular service accounts for almost all telephone subscriptions – for example, Congo (97.9 percent cellular), Uganda (92.7 percent cellular) and Cameroon (90.1 percent cellular).

Internet access and use is also a major element of the knowledge economy infrastructure. In 2003, the ITU reports over 219 million hosts, or computers linked to the Internet, supporting almost 700 million Internet users. Data on Internet hosts by country are presented in Figure 9. The pattern is far more concentrated than telephony data, showing most hosts located in North America, Europe, Japan, South Korea, Australia and Latin America. Worldwide, there are 357.7 hosts per 10,000 inhabitants, with the highest levels in the United States (5577.8 hosts), Iceland (3789.7 hosts), several countries with over 2000 hosts (the Netherlands, Sweden, Denmark), and those with approximately 1000–1500 hosts per 10,000 population (Australia, Canada, New Zealand, Japan, Singapore, Taiwan). In many poor countries, there are fewer than 1000 hosts for all residents – in some cases less than one computer connected to the Internet per 100,000 people.

Another perspective on Internet distribution and use is the number of Internet users globally, presented in Figure 10. The worldwide population of Internet users in 2003 is estimated at almost 700 million people, or 11.4 percent of the world's population. Users as a percentage of population by world region ranges from a high in Oceania (43.0 percent) followed by the Americas (26.4 percent), Europe (24.2 percent), Asia (6.9 percent), and Africa (1.56 percent). The pattern of users differs from the distribution of hosts, with a number of

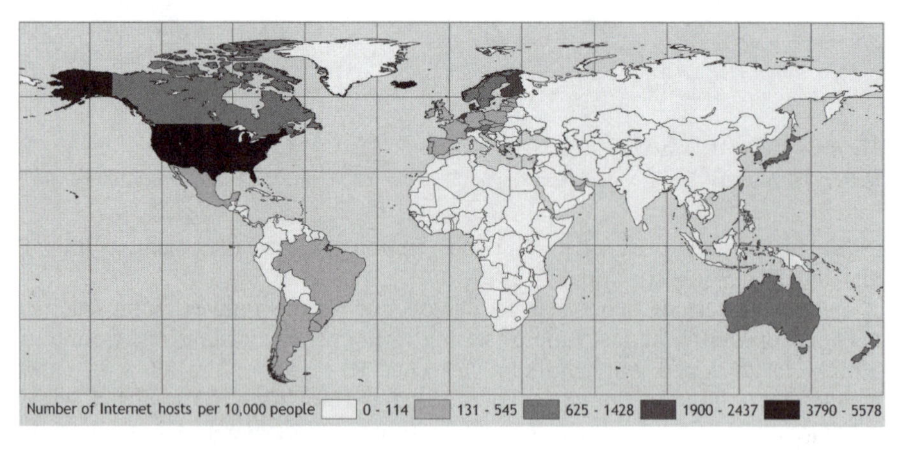

Figure 9 *Internet Hosts (2003)*

Source: International Telecommunication Union (2005), Internet indicators: hosts, users and number of PCs.

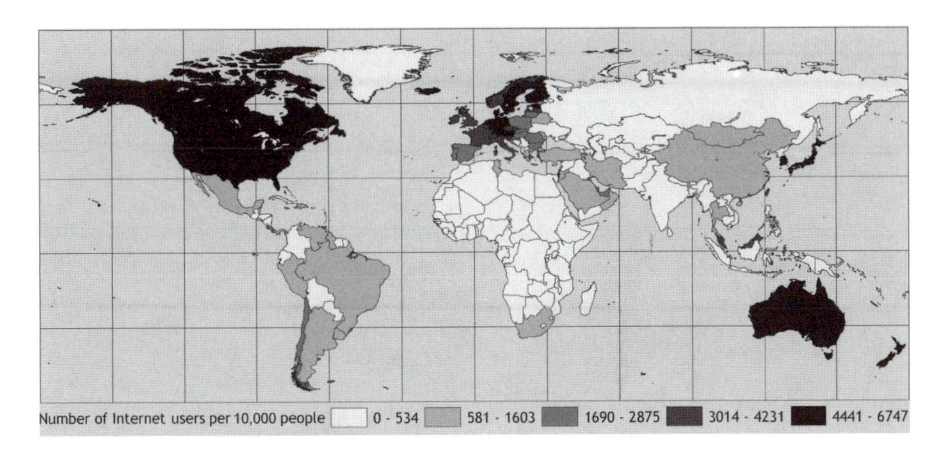

Figure 10 *Internet Users (2003)*
Source: International Telecommunication Union (2005), Internet indicators: hosts, users and number of PCs.

developing countries having fewer hosts but many users. For example, fewer hosts/many users applies in China, Thailand, Iran, Saudi Arabia, Turkey and South Africa.

Access and use of the Internet can be, and should be, analyzed at many levels. For example, a study of Internet use in Michigan in 2004 shows variation across regions in frequency of use, presented in Figure 11.

Responding to ICT

Different locations have different experiences and reactions to the knowledge economy and ICT infrastructure development. Local responsiveness to new technologies and economic change reflects economic conditions, innovation, and political and community leadership. Patterns of response to ICT vary for many spatial layers, such as globally, nationally, within states and provinces, and even within metropolitan areas.

A summary of country variation in ICT preparedness is published by the Economist Intelligence Unit (2005), with e-readiness rankings based on an index of technical, business, political and social measures. The 2005 rankings for world regions are presented in Figure 12, which also shows high and low ranges for each region. The leading regions are North America and Western Europe, with high scores for countries such as Australia, Singapore, Taiwan, Hong Kong, South Korea and Japan. Characteristics that are evident for leading nations include policy commitments to ICT development, such as broadband rollout, as well as firm innovation, citizen adoption and use of ICT, and education.

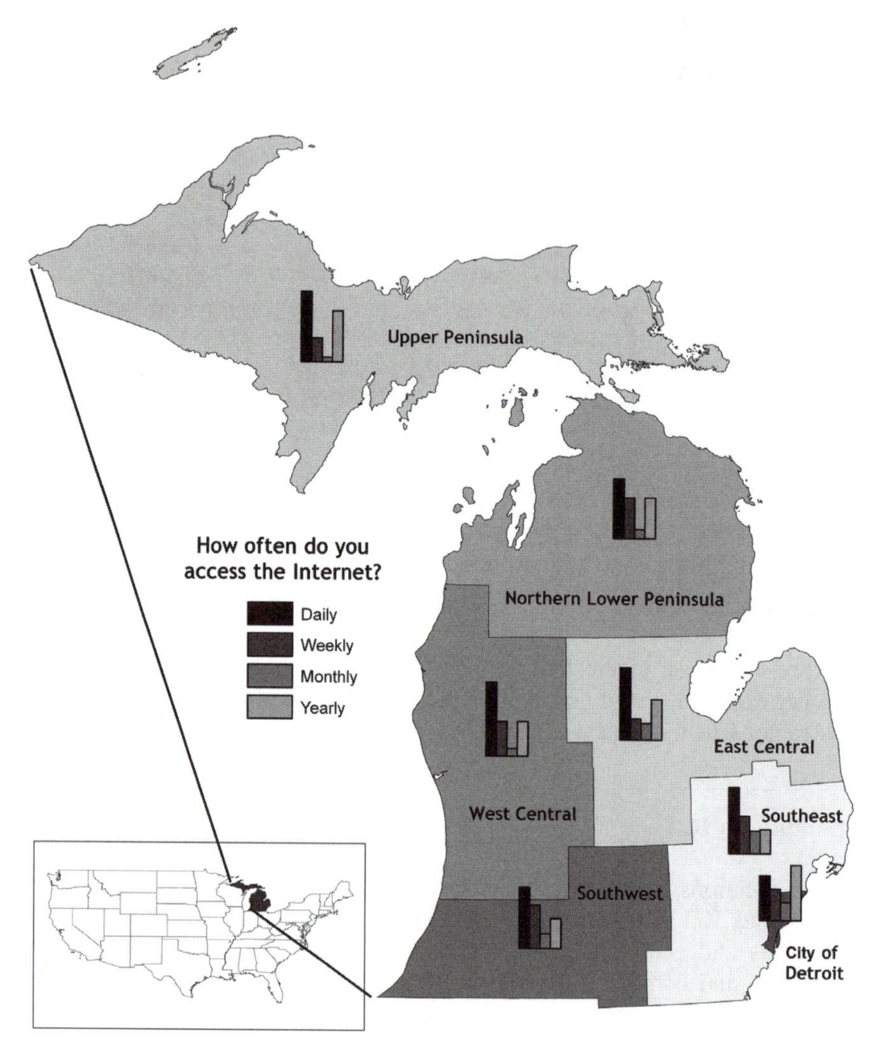

Figure 11 *Internet Frequency in Michigan (2004)*
Source: Wilson, M.I., Corey, K.E., Helmholdt, N. and Frederick, E. (2004).

A similar study was conducted for states (2002) and cities (2001) in the United States by the Progressive Policy Institute (PPI). The 2002 State New Economy Index identifies leaders and laggards among American states in terms of their economic dynamism and uptake of ICT. Among the leading states were Massachusetts (90 out of a possible 100), Washington (86.2), California (85.5) and Colorado (84.3), with lagging states of West Virginia (40.7), Mississippi (40.9) and Arkansas (41.7). A similar index was calculated by PPI (2001) for

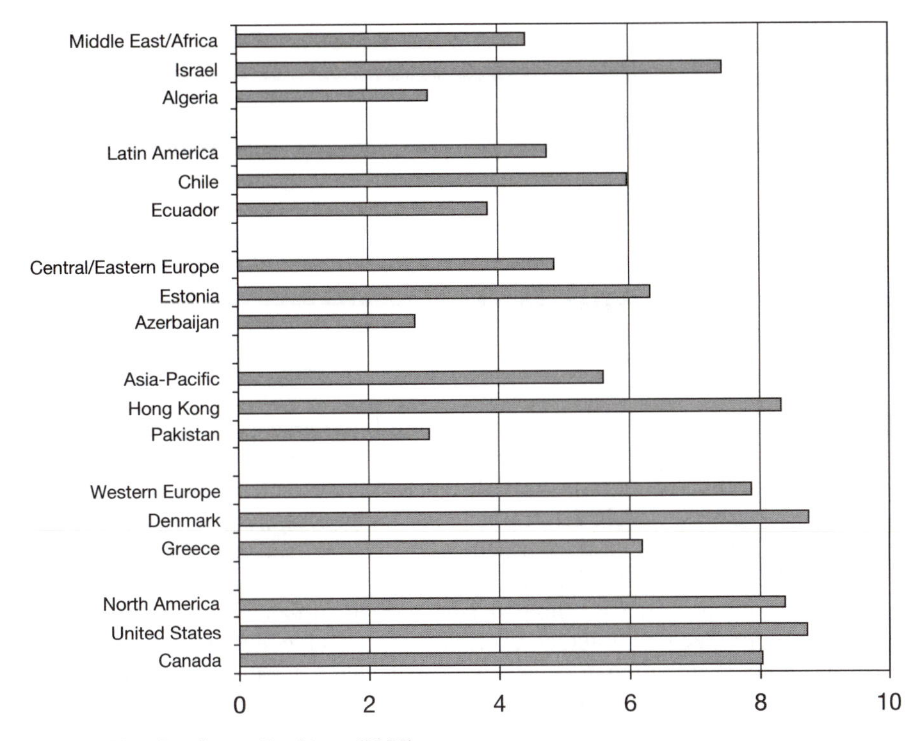

Figure 12 *e-Readiness Rankings (2005)*

Source: Economist Intelligence Unit, 2005 e-readiness rankings.

metropolitan areas, with the leading location being San Francisco with a score of 95.6 out of a possible 100, followed by Austin (77.9) and Seattle (68.0).

Moving to a lower geographic layer of analysis, the new economy index was applied to the counties of Michigan (LaMore *et al.* July 2004), showing considerable variation within one state. Counties were identified as leaders, followers, contenders, and laggards, with some remote counties serving as leaders, while some counties close to major metropolitan areas lagged behind the state average. A map of Michigan's knowledge economy index is discussed further in Part III.

Place, Space and ICT

The power of ICT to conquer distance, using bits rather than atoms as noted by Negroponte, changes the meaning of distance and also space and place. The ability to interact across vast distances easily and at low cost led to many spatial analyses of ICT's implications for geography. Hepworth (1990), Kellerman (1993 and 2002), Cairncross (1997), Wilson and Corey (2000), Dodge and Kitchin (2000; 2001), and Zook (2005) all have contributed to our understanding of the meaning of space in a virtual world. With cities a product of our need to interact closely for economic and social purposes, analysts directed attention to how cities will change. Cities as a cause and effect of industrialization prompt questions about population concentrations in a post-industrial setting. The future of cities was examined by Castells (1989) and Drennan (2002), with urban life challenged by Mitchell (1995; 1999), and caution from Graham and Marvin (1996; 2001) about the costs, as well as the benefits of ICT mediated urban life.

The death of distance, proclaimed by *The Economist* in 1995 and later elaborated by Cairncross (1997), captures the sense that distance no longer matters. The conquering of distance in electronic terms brought for many the dream of low-cost and efficient access to people and information worldwide. The death of distance, however, is too simplistic a claim because it misses so many of the nuances of the structure and character of electronic interaction. Distance may not matter for those in affluent countries, or working in information industries, or for those with business or personal reasons to contact others. Those for whom distance may not matter are a small proportion of the world's population. For many of the world's inhabitants, distance remains a source of social and economic friction, and place matters more than ever.

Often associated with the concept of the death of distance is the mistaken assumption that it also refers to the declining importance of place. Low-cost electronic interaction has certainly made it easier for residents of different places to interact, but at the same time it has also brought different places into a common realm where differences matter. Rather than be seen as a force diminishing distance, low-cost electronic interaction underscores the powerful value of connecting places. The reason that so much effort and investment has been directed at developing electronic infrastructure is because of the value gained from interacting with different places. There are many reasons for wanting to access different places, such as their markets, the information produced or exchanged there, sale of information as an input to production, or to connect people who have family ties across space.

Globalization as a process was well developed for manufacturing, and the worldwide supply chains of industry was one incentive for the development of transportation infrastructure and its accompanying coordinating function of information and communication technology. The patterns of globalization seen for manufacturing were thought inapplicable to services, yet by the late 1980s the offshore back-office phenomenon, and the later outsourcing of business and financial services showed how even the advanced end of the service economy was not immune to relocation to lower cost and well-connected cities and regions.

Observing the global processes of production and information flows, Sassen (1991; 1994) recognized the rise of a few major metropolitan areas that were central to the circuits of globalization. New York, London and Tokyo emerged as global cities where the command-and-control functions of the world economy were concentrated, supported by a second tier of cities such as Paris, Chicago, Frankfurt, Hong Kong, Los Angeles, Milan, Singapore, San Francisco, Sydney, Toronto, Zurich, Brussels, Madrid, Mexico City, São Paulo, Moscow and Seoul (GaWC (Globalization and World Cities Study Group) n.d.).

Issues for the Knowledge Economy

Analysis of the knowledge economy shows the significance of information and knowledge as core functions in an advanced economy. Economic success and wealth accrue to those associated with the knowledge generation functions of the economy, whether they are skilled and educated workers, owners of capital invested in knowledge firms and industries, or those who serve the economic needs of the knowledge sector. What then is the forecast for communities that do not contain knowledge firms and industries, and whose residents do not have access to the education resources essential for employment success?

Before continuing, it is valuable to identify some of the divisions that are associated with the knowledge economy. First, there will be economic divisions between industries and workers that form the core and periphery of the knowledge economy. Communities with robust knowledge-based industries and advanced services will provide growing incomes and quality of life for residents, while areas losing manufacturing and having little to replace it will find fewer opportunities for growth. Workers with higher education or technical training will be able to maintain their living standards, while those with less education will struggle to keep up.

A second division will be based on geography, with cities and regions that contain clusters and agglomerations of knowledge production, contrasting locations that cannot compete in a knowledge economy. Expanding consideration beyond the nation, we can consider the divisions between countries that gain or lose in a knowledge environment. Layers of have and have-not regions will emerge, with polarization at the metropolitan, regional, state/province, national and international levels of geography.

Social divisions represent a third gap, between owners of knowledge capital and their workers, and those whose capital is not recognized as economically valuable. Knowledge workers will face growing economic opportunities, and will tend to cluster in metropolitan areas of relative affluence (Graham and Marvin 1996). A final division will be political, between the interests of the knowledge economy and those representing the interests of excluded or peripheral participants.

A great deal of attention in planning is oriented to finding the sources of growth or ways to capture creativity and knowledge. The converse challenge is to deal with those places and communities that are not competitive. Communities may be distressed in a number of ways, such as lacking education resources and support for residents, or having no economic link to the growth industries of the day. A community may be in very close proximity to knowledge work and firms, yet be unable to benefit because its residents lack the skills and training needed, and are excluded from this source of economic growth. Distress stems from the lack of training or preparation for the jobs in demand. As the premium paid to knowledge workers increases, those without sufficient educational preparation slip farther behind. The most dramatic example of this trend is the decline in real wages over the past two decades of workers without a college education.

Conclusion

As the economy changes and places greater emphasis on knowledge work and employment, communities will face varying experiences. Some will benefit greatly due to their well-educated workforce and proximity to the centers of innovation, and research and development for the global economy. Other communities will face challenges finding a path that will offer opportunity to workers who may have received less training and education, or for communities in distress from the deindustrialization process. The positive future foretold by Daniel Bell may have come true for educated workers in a post-industrial society, but for the people and places disconnected from the growth sources, the future is less positive.

The purpose of this part of the book is to identify some of the changes that contemporary planners need to understand and then use. Planners face many new conditions in cities and regions, including (1) changes in production, with the shift away from manufacturing to services, and also toward advanced services and knowledge work; (2) the impact of information and communication technologies on how and where services are produced and how people interact; (3) implications for social justice of uneven access to ICT; and (4) the potential change associated with ICT and globalization. The first task, then, is to develop awareness of the forces affecting our cities and regions, and then to use these forces for the development of our communities.

PART II
Concepts

Key Concepts and Their Roots

This part of the book lays the conceptual groundwork for constructing a model for effective planning practice in this era of the global knowledge economy and network society. We reiterate that this process and the resulting model are our examples for the several contexts that we have experienced. Our examples are intended to stimulate your own modeling and local tailoring for your city-region. In being informed by these examples, we would expect you to draw on your own selection of concepts and theories that are most applicable and appropriate to the particular development circumstances of your locality.

From Digital Development to Intelligent Development

With these and other emerging relational conceptualizations, today's regional planning practitioners may move beyond mere *digital development*, which is focused on ICT access, to *intelligent development* that is informed by appropriate theory – i.e. relational theory, sound empirical analyses – and thereby realize more effective contemporary planning practice and development outcomes.

Regional planning practitioners and scholars will continue to learn from experimentation and innovative practice with the new development opportunities that are offered by addressing the full range of these and other relational approaches. In future, promising additions to this emerging body of relational theory-informing planning practice may include the integration of actor network theory, more attention to the division of labor inherent in the stage-based roles, the elaboration of actors, the roles and agency inherent in the Program Planning Model and its organizational relation – i.e. the Program Management structure for innovation generation – and equity planning theories, among others. In addition to the need for planning practitioners to engage relational approaches, it is believed here that planning theory also can benefit by having planning theorists actively take up the many challenges of advancing relational approaches as part of the general body of evolving planning theory.

Digital Development

Digital development is the provision of, and access to, information and communications technologies (ICTs) infrastructure for community economic development. Among others, high-speed broadband networks and wireless systems are examples of digital development infrastructure. Even in developed economies, there are far too many places and regions of the world that are not served by ICT infrastructure. However, the roll-out of digital development infrastructure is underway and expanding. While this distribution proceeds, the critical strategic planning issue is to prepare for and develop content for application by means of the ICT infrastructure. The greatest need is to plan creatively and effectively for content application. This means putting high priority on "intelligent development," while the digital development continues.

Intelligent Development

What is intelligent development? It is digital development that explicitly draws on and is guided by theory, especially location theory, economic development theory and relational planning theory. Selections from these bodies of theory can be instrumental in the formulation of policy and the planning of communities for effectively competing in today's global knowledge economy and network society (Graham and Marvin 2001; Castells 1996). Intelligent development promotes investment in a place, thereby producing wealth creation, human capital development, employment formation, creation of an enterprise culture and improvement in the region's quality of life (cf. Komninos 2002; Droege 1997). Such planning is focused on maximizing the value-added of a place by matching and segmenting the unique e-business functions, factors and relations of the locality and its region to the appropriate stages in the life-cycle processes of production and consumption functions. Development is "intelligent," therefore, when the best practices from theory, from benchmarking elsewhere, and from appropriate applications of the latest technologies are utilized fully to develop a community holistically, multidimensionally and equitably. This theory-informed approach to planning and plan-implementation has been conceptualized as "intelligent development" (Corey December 19, 2003). Development is "intelligent," therefore, when ICTs are an explicit and facilitative part of the planned development strategy. Intelligent development is stimulated by best-practice cases, by successful testing of development and planning theory, from benchmarking comparable city-region cases elsewhere, monitoring and evaluating the status of plan-implementation, and from appropriate applications of the latest state-of-the art technologies.

Background for Relational Conceptualizing – The Legacy of Jean Gottmann

A generation ago, in the 1970s, when we were seeking to do planning research on long-range planning for the future of the metropolitan region of Seoul, Korea, we searched for appropriate planning theory and concepts that might inform our planning-practice approach. Our search revealed little that was helpful, with one exception. That exception was the rich body of work – i.e. the extensive writings and interpretations – of French geographer, Jean Gottmann.

Since the publication of his 1961 landmark book, *Megalopolis*, Gottmann had continued his observations of the stimuli and drivers of the new urbanization processes that he had identified as emerging throughout the northeastern seaboard of the United States in the 1950s. He attributed these new economic and development drivers to "quaternary economic activities" and "white-collar occupations" that were operating to transform advanced services and the landscapes that housed these growing service sectors. Until his death in 1994, Jean Gottmann published and extended his pioneering research into the emerging urban-regional dynamics of the megalopolitan and transactive forces that are driving the global knowledge economy and network society of today. For definitions and references to these various terms and concepts, refer to Part V, section entitled Gottmann Concepts, A–Z, pp. 221–7.

Gottmann's foresightful contributions were helpful to our early work a generation ago, and it continues to be informative for us to this day. Since the time when we searched for conceptual guidance a generation ago, the theoretical inspiration for our work has deepened and widened considerably. Today, there are many new concepts and contemporary research findings upon which to draw. We have found that *relational planning* is a body of ideas and reflections that is quite compatible with the complexities and nuances of today's technology-enabled economies and societies. Stephen Graham and Patsy Healey's 1999 important article on relational concepts for planning theory and practice is a noteworthy case that makes this point. In addition to these new ideas, Gottmann's legacy of earlier compatible ideas continues to complement and advance the goals of this book. Without necessarily using the label, his concepts and approach were "relational" in nature.

Early on in the emergent stage of the global knowledge economy, Jean Gottmann noted the criticality of "people relations" as being at the core of the white-collar revolution (Gottmann 1961: 627). This was a prescient observation from an astute, and one of the earliest students of the roots of the global knowledge economy and network society. It foretold of the complex and interdependent world of today's and tomorrow's globalization. Professor Gottmann's observation about people relations foretold also of the development of relational theory and its application to urban and regional planning and development. Below and throughout the book, we elaborate on the criticality of relational theory to the practice of regional and local planning in the context of the global knowledge economy and network society.

For all the value of so much of this work and concepts, the practice of planning and the profession of planning around the world remain challenged and require our collective attention and creativity to meet these new demands. Next, we reference a selection of constructive suggestions from Sanyal and Markusen. Their comments are used here to stimulate and focus the construction of our model for regional and local planning practice in the global knowledge economy and network society.

Challenges

Challenges for the Practice of Planning Today and Tomorrow

In reflecting on these issues, planning scholar, Bishwapriya Sanyal has concluded that the profession of planning faces some major challenges. Consequently, the profession needs to: (1) integrate spatial and socioeconomic planning; (2) construct planning theories to meet the needs of planning practitioners; and (3) rejustify government intervention (Sanyal 2000: 317–33). The model for planning practitioners constructed for this book is intended to enable local planning practitioners to begin to address these challenges by inventing their own planned solutions. To meet and operationalize these challenges, planners will need to apply the principles and lessons from the emerging and growing body of contemporary planning theory and practice. Further application and use of these theories will serve to advance the effectiveness of the practice of relational planning at the regional and local levels, such that it becomes increasingly integral to and engages the dynamics and complexities of the global knowledge economy and the network society.

"One reason for the relatively poor image of the profession is its lack of ideas" (Sanyal 2000: 318). It is intended that the ideas and constructs introduced here might serve as stimulants for local planners to devise their own relational planning approaches to advance the economic competitiveness and quality of life of their planning region. As a result, this stimulation may be expected to generate a profusion of ideas and innovation, thereby rendering the profession more effective and societally useful.

The Challenge of the Need for the Integration of Spatial and Socio-economic Planning

Among several of the other conceptual constructs introduced in this part of the book, the e-Business Spectrum, described below, seeks directly to address this challenge. It can be a useful construct for planners, because it has identified the dominant spatial and locational patterns that are associated with the principal economic and social functions of regions and localities, especially in the context of the global knowledge economy. It can be used further to stimulate the construction of prescriptive and normative strategies that are congruent both with the spatial patterns and to the development needs and empirical realities of the particular planning region and its localities.

The Challenge of the Need for Planning Theories to Meet the Needs of Planning Practitioners

There is a growing body of rich and diverse planning theory that was stimulated by the new and dynamic circumstances of the global knowledge economy. Relational theory, and its principles, represents promising creative stimuli for the local regional planner who is engaging these uncertainties and complexities. As portrayed here, a number of related conceptual constructs have been introduced or used to explicate some ways that these concepts can be useful in planning practice.

The relational concepts noted above include reference to Graham and Healey's (1999) relational planning practice framework (October 1999). That framework consists of four elements: (1) *Relations and processes*: in the book here, this is operationalized as the functions, factors and principal segments of the e-Business Spectrum elaborated below. (2) *Space and time*: in addition to the spatial organizational patterns of concentration and dispersion of the production, consumption and amenities functions, the *space* part of this element is illustrated further in the form of the six locational and economic development theories that have been reviewed and tested by Plummer and Taylor (2001a; 2001b; 2003), and a composite construct that integrated measures of these six theories is derived. The *time* part of this element is illustrated by means of the Program Planning Model (Van de Ven and Koenig, Jr Spring–Summer 1976). It is a planned change process that functions over time, that incorporates relational propositions, especially by differentiating planning phases by the role of particular actors and by their principal contributions by stage or phase to the overall planned change process; the importance of the "moment" in time and in planning also is introduced (Sull July–August 1999; 2003). (3) *Multiple layers of power geometries by place*: these may be depicted empirically as place-specific layers of mapped knowledge-economy functions and their respective spatial distributions. (4) *The power of agency in negotiating between the layers of power geometries*: this relational planning element may be operationalized by defining agency empirically and then analyzing the linkages (or needed linkages) among the layers, especially with the intellectual support, guidance and application of the appropriate theory and concepts.

The Challenge of the Need to Rejustify Government Intervention

As elections result in shifts in governments and changes in regimes, different development-policy directions may be taken by the new government and new regime. Associated with these changes, public attitudes too may shift, such that belief in market-led development in contrast to government-led interventions for development may wax and wane over time. These different approaches to development are not necessessarily in opposition. In today's global knowledge economy and network society, what is likely to be most effective for ICT-enabled regional development planning is complementarity among a locality's governments, the market, the business sector, the nonprofit sector and its institutions and individuals. Working in concert, each of these sectors has an important contribution to make to regional and local development, especially over the phases or life-cycle of a strategic development-planning process. This mix of roles played by the respective key development actors is likely to be effective because this approach is representative of the wide and diverse base of a locality's principal development stakeholders and their assets.

Given the inherent resistance and reluctance to formally reorganize and restructure government, the more likely early response locally is the formation of governance groups for the region. Along with other local actors, formal government representation from various levels of government thereby may contribute as important partners in planning and implementing regional development strategies. To be most effective, the governance approach requires high levels of coordination and institutional complementarity. Such governance effort should benefit from the leadership and services of professional urban and regional planners, who may function in local leadership and support roles. In places with few, if any, planners, others may assume such roles; often economic development professionals and other citizen planners have demonstrated their capacity to advance the planning and execution of the plans.

The Challenge of Creating a New Mindset for Planning

Sanyal concluded his discussion of the challenges of the planning profession by stating that the most important challenge is to create a mindset that

> require[s] strong commitment to social progress and a worldview that government, market and civil society must complement each other in moving forward toward that goal. How to create such a mindset, not only among planners but among all citizens, remains the single most important challenge for the planning profession.
>
> (Sanyal 2000: 332)

Refer to Mindset in Part V, section on Relational Planning Concepts, A–Z, p. 210. It is the predicate of this book that relational planning practice can serve as the beginning of a process that can result in the development of a new mindset for planning practice. This needed mindset, in time, will have to become embedded in the routine and daily behavior of planners and their stakeholders. Core to the new mindset must be the full incorporation of ICT-enabled development that drives creative production, consumption, and amenities, and quality of life factors as embodied in the e-Business Spectrum introduced below, and as informed by relational planning theory and its guiding principles, including the mobilization of current and future community assets to address the needs of economically distressed communities.

Empirical assessment of relational planning efforts has demonstrated that strong inertial forces, such as embedded governance rigidities, work unconsciously to thwart change in the practice of planning and in organizations and institutions that are stakeholders in and influence the regional systems of planning – i.e. statutory and institutional environments. These observations are reinforced by path dependence and legacy approaches from the past. Critical, therefore, for the planners and stakeholders of the region, in a paraphrasing of Donald Sull's concepts, is the need *to create and sustain a culture of institutional consciousness* (Sull July–August 1999; 2003). The need to change the planning mindset needs to be explicit and continuously reinforced in everyday planning practice. Ultimately, however, the key elements of the practice of relational planning will require the mandate of law and formal statutory regulation – e.g. the coordination or integration of telecommunications infrastructure and physical infrastructure as part of the official or legal brief of the region's public planners' agencies. Such actions would be intended to normalize and routinize the practice and implementation of new relational planning. To be responsive to the dynamics of the global knowledge economy and the network society, new organizational and behavioral environments for planning need to function in flexible and agile ways. For example, it is important for practicing relational planners to identify and act on these barriers to innovation and to be proactive and responsive to the constantly changing environments that are external and internal to the planning region. By creating such prerequisites, promoting the convergence of supportive episodes and by *waiting* for the right moment, planners, their leaders, and stakeholders may be positioned to make more effective use of time and the timing of strategic interventions. Refer to Mindset in Part V, section on Relational Planning Concepts A–Z, p. 210.

Concept to Action

Actions for Planners to Take

Much of the discussion referenced here by Sanyal is concerned with enabling professional planners to fulfill their roles in society by being effective in addressing today's challenges. It is the premise in this book that planners must change their mindset. Further, they need to work to ensure that the stakeholders and the region's localities develop mindsets that also are compatible with the new requirements of the global knowledge economy and network society.

To operationalize these premises, Ann Markusen has proposed three approaches for ensuring that planning may enhance its relevance and centrality to today's new development requirements: "planners must engage in public discourse to influence public opinion; they must showcase planning's 'best practices;' and they must diversify into new fields" (Markusen 2000). This part of the book's discussion has sought to point toward operationalizing the new field of relational planning. It is hoped that Markusen's other two approaches of public discourse and the showcasing of new planning's best practices will be taken up and implemented by planners throughout the profession. Our treatment of these issues in Part IV is intended to offer examples of how local planners might operationalize the challenges and premises discussed.

Conceptual and Theoretical Frameworks for New Planning Practice

Until recently, there was little in the way of theory that practicing regional and local planners could use to inform their global knowledge-economy planning strategies. Today, however, this has begun to improve with emerging conceptual frameworks that should be tested (Graham and Healey October 1999; cf. Rodwin and Sanyal 2000). Relational theory especially is promising. See the definition of relational theory in the section on Relational Planning Concepts, A–Z, in Part V, pp. 214–16.

Relational thinking and theorizing are congruent with the complex, multi-layered functions and flows of today's formal economy and networked society; refer to network society in Part V, section on Relational Planning Concepts, A–Z, p. 211. Thinking in relationality terms can serve to stimulate our imagination, and free up our collective and conventional neoclassical economic and locational perspectives of the world and our legacy approach to planning. For example, Henry Wai-chung Yeung has discussed the "explanatory power in socio-spatial relations among such actors as individuals, firms, institutions . . . and other nonhuman actants" (Yeung March 13, 2002: 2).

In reworking his earlier conceptualizations of relational economic geography, Yeung has extended and clarified his formulation of relational theory (Yeung 2005). As a result of this refinement, he has identified and clarified three sets of conceptual connections that might stimulate relational thinking and behavior within the context of economic space. First, there is actor-structure relationality. Second, the relationality of scale may be conceptualized by layers that are global, national, regional, and local. Third, relationality may be perceived as the connectivities between, economic, social, political and spatial elements (Yeung 2005: 43–44; cf. Castells 1996; May 9, 2001).

Adopting relational conceptualizations, Stephen Graham and Patsy Healey (1999) have provided planners with useful initial theoretical framing for incorporating the new technology-based geography into visioning and planning for development. Based on relational theories of economic, societal and cultural functions, time and space, and power dynamics, they argue for needed changes in planning practice. Four interrelated guiding points are suggested for the practice of relational planning; they suggest that we: (1) consider *relations and processes* rather than only objects and forms; (2) stress the multiple meanings of *space and time*; (3) represent places as *multiple layers* of relational assets and resources, which generate a distinctive power geometry of places; and (4) recognize how the relations within and between the *layers of the power geometries of place* are actively negotiated by the power of agency through communication and interpretation. Refer to Part V, section on Agency, p. 189. Each of these guiding points, or relational planning elements, is elaborated below.

Relations and Processes: The e-Business Spectrum as a Functionally-based Organizing Framework

Relational theory is the primary conceptual context for the electronic-business spectrum, which is described next. From here onwards in the book, this construct is referred to as the e-Business Spectrum. Relational theory is useful here because it does not put sharp edges on concepts (Yeung March 13, 2002). Rather, it accommodates and explains relationships in realistic terms – i.e. terms that reflect well the empirical complexities of today's networked and connected world. For example, when regional leaders and planners construct planned visions for their region of responsibility, it is imperative that their planning for "infrastructure" fully integrates holistically high-speed, high-capacity ICT infrastructure into the region's overall infrastructure planning system. From a global knowledge economy competitive context, substate regional planning cannot afford to continue to attend *only* to such pre-information age physical infrastructure as transportation; sewers; water supply; electrical power, and other tangible utilities. Some actor in society needs to assume the critical strategic role of integrating the new electronic infrastructures into the everyday context of the planners' traditional mindset of planning for and controlling the development of the mechanical and physical infrastructures of local communities. Indeed, "knowledge economy infrastructure" requires full planning attention also to include education and higher education infrastructure (Dowall and Whittington 2003), as well as to include ICTs or cyber infrastructure, in the regional planning that must be practiced today and for tomorrow. It is imperative for the new planning mindset to incorporate fully these new digital development technologies into regional and local planning. Sooner, rather than later, these new factors need to go beyond mindset change and be statutorily mandated as a routine part of the local planner's mandate.

Given the functional complexities and the expanding pervasiveness of the ICT infrastructure of the information age, global knowledge economy, and network society, it is useful to have a functional and relational framework to assist in ordering our thinking about at least some of these complexities. The e-Business Spectrum has been constructed to address this need. This construct stands for, and is shorthand for the full range of the generic and specific "business of electronic-related and driven economic development." For short, it is referred to as the e-Business Spectrum. It spans the principal economic development functions of regional and community analysis, planning and development implementation. These functions include three overarching ICT-facilitated and knowledge-economy activities that are integral to today's digitally facilitated work and living environments: (1) production functions; (2) consumption functions; and (3) amenity and quality-of-life factors. The e-Business Spectrum is congruent with Henry Yeung's reconceptualization of the nature of relationality within the context of the formal economy (Yeung 2005). The e-Business Spectrum is a conceptual construct of economic functions, including the linkages, and social and political power and spatial relationships among those functions. These relations function at various layers or scales.

The e-Business Spectrum was conceived with the inherent bias that the production functions are the region's basic economic drivers with the other functions following as dependent variables. It should be noted that all these functions and factors lend themselves to innovation, creativity, and informatization; these characteristics are not the sole province of computer hardware and software industries. The science and technology, and the arts and culture

The "Business" of Electronic Driven Economic Development		
Production functions	Consumption (e-commerce) functions	Amenity and quality-of-life factors
Science and technology-driven research and development (C)	Online procurement: B2B and B2G and G2G (D)	Innovative social, cultural and institutional activities (C and D)
Commercialization of products and services (C)	Online retailing: B2C and G2C (D)	Natural environmental attributes (C)
Business and producer services (C) and manufactured products (D) Public and government producer services (C and D), e.g. regulations, taxes, info., etc.	Value-added complementarities between electronic (clicks) and physical (bricks) channels (C and D)	High-quality education and human capital capacity-building and talent development (C and D)

Figure 13 *e-Business Spectrum (Conceptual Framework)*
Source: Authors.

components of the e-Business Spectrum especially are instrumental in contributing to the development of a climate of creativity for growing a regional and local culture of innovation and entrepreneurship. Each of these domains of the e-Business Spectrum has its own dominant *locational and spatial organizational pattern*. These general geographical patterns are symbolized in Figure 13 as (C) for concentrated or clustered patterns and (D) for dispersed or distributed patterns (cf. Kellerman 2002). Note, given the popularity and wealth of written material that is readily available on "clusters," we will not elaborate here on this important concept; refer to Part V, section on Clusters, pp. 194–5. The e-business framework presented here can be used as both an analysis tool and as a planning organizing tool (see Figure 13).

Production Functions

Places that value and assign priority to creativity and innovation are behaving in compatible ways to be competitive in the global knowledge economy. Emphases on science and technology, and the arts and culture can operate to reinforce the creation and sustainability of a local creative culture. Charles Landry's book, *The Creative City: A Toolkit for Urban Innovators* (2000), offers helpful insights into these dynamics. Also refer to Richard Florida's *Cities and the Creative Class* (2005a). The front end of the innovation and creativity process, especially science and technology-driven research and development, frequently is based on tacit knowledge. Refer to Industry Life-cycle Model in Part V, section on Relational Planning Concepts, A–Z, pp. 203–4. Consequently, these often face-to-face communications require proximity. The resultant locational pattern is one of clustering or concentration. This is the phase during which the initial development of intellectual property is likely to occur. See Margaret Pugh O'Mara's book *Cities of Knowledge* (2005). See also

Cities of Business Leadership and Enterprise in Part V, section on Relational Planning Concepts, A–Z, p. 193. A later phase of the industrial products and services life-cycle – i.e. commercialization of products and services – also functions most effectively in a face-to-face, tacit knowledge communication and spatially clustered context (cf. Reamer 2003). Part of the commercialization of intellectual property may require venture capital development. This, too, typically has functioned most effectively with proximity between an innovative invention and the source of venture capital (cf. Zook 2005). The spatial distribution of venture capital opportunities varies widely from region to region. As business and producer services need to be delivered in close proximity, these activities should be concentrated geographically. However, the manufacturing of business and producer products can cluster, but they also can locate in distributed or dispersed locational patterns. It should be noted that sometimes the label "post-industrial" is used in discussions of high-technology production and economic sectors; this does not mean that manufacturing goes away or is not part of the policy and planning context of the global knowledge economy and network society; rather, it means the era in the evolution of an economy has been reached, after which manufacturing no longer is the dominant means of production, as in the case of an economy where Jean Gottmann's quaternary economic activities have assumed dominance locally. We make this point to stress that places that have been manufacturing centers may plan to maintain this means of production as an important sector, even as the actors of the region also seek to thrive in the context of the global knowledge economy and network society. For an elaboration of Quaternary Economic Activities, refer to Part V, section on Gottmann Concepts, A–Z, pp. 226–7.

Consumption Functions

Depending especially on the extent, density of, and access to ICT infrastructure and spatial distribution of such connectivities, electronic commerce activities of online retailing and online procurement can occur nearly anywhere. Consequently, the dominant locational pattern is one of dispersion. Planners may support the advancement of electronic commerce consumption functions by using the e-commerce planning elements that have been identified as key to successful e-commerce locally and regionally. These e-commerce planning elements include: (1) vision and leadership; (2) modern ICT infrastructure; (3) financing and investment; (4) regulatory environment; (5) human resources and training; (6) culture and political economy; (7) multiple sectoral applications – e.g. manufacturing, retailing, e-government; (8) local and regional applications and cases – e.g. the benefits of integrated physical and online commerce such as "bricks and clicks" business models (Steinfield 2003); (9) organizational behaviors and dynamics; (10) timing; (11) spatial organization; (12) research and evaluation; and (13) self-learning and learning resources – e.g. websites and maintenance of a purpose-constructed "e-commerce primer" for the locality (Corey May 2002; Fletcher *et al.* 2001).

Amenity and Quality-of-Life Factors

Many years ago, the geographer Jean Gottmann noted the importance of the "hosting environment" (refer to Part V below, section on Gottmann Concepts, A–Z, p. 224), wherein "the quality of work and the quality of life" are critical factors in enhancing the attractiveness of a location; this is true especially for the knowledge worker and her/his family in the information age and global knowledge economy of today (Gottmann January/February 1979: 5). The quality of life at a location can be enhanced because of the attractiveness of a community's social, cultural, institutional and natural environments. The quality of education and the educational and human capital capacity of the community and its region are critical (Plummer and Taylor 2003). Successful regional labor markets must have sufficient streams of talent to attract and retain employers and job creators. Families, in general, and the families of knowledge workers in particular demand a high-quality local educational system. This system must demonstrate that it has the capacity to generate and sustain a regional workforce with the talent capable of nurturing an ICT-based and knowledge-driven local economy, along with its pre-existing traditional economic capacities – e.g. manufacturing production functions.

There are practical concerns that need to be taken into account when planning for human capital as a high-priority investment. Susan Clarke and Gary Gaile have labeled one of the principal concerns, "the leaky bucket dilemma." They have written:

> The logistical difficulty of capturing returns complicates investment in developing human capital. The mobility of human capital, its flexibility, its 'non-ownability' and the long time lags involved before returns are realized serve to discourage local public and private investors.
>
> (Clarke and Gaile 1998: 193)

In order to retain and attract young adults to its cities, the government of the State of Michigan has implemented the Cool Cities program. For an elaboration of the program, refer to "What is Cool Cities?" (Michigancoolcities.com n.d.). Similar to many areas around the globe, Michigan and its communities have invested heavily in education, and long-term and significant investment in tertiary university-level education, including major research universities. More of its recent university graduates leave the state than is healthy and productive for its local economies. Over time, it will be informative to follow the status and results of this initiative. As technology-based and science-enabled economic development strategies become increasingly pervasive, the push-and-pull forces associated with quality of life and amenities factors will continue to play critical roles in regional and local development planning. It is imperative that each locality should make human capital investment a high priority, while simultaneously inventing and implementing effective ways that tap into the reasons why knowledge workers decide to stay or move from the places that have invested so importantly in their human-capital development. For the example of the Michigan Cool Cites Survey, refer to (Michigancoolcities.com n.d.).

Attention to, and priority investment in the full range of requisite high-performance educational institutions in the local region should include: early childhood educational programs; primary and secondary schools; polytechnics and community colleges; and universities – including articulated access to and commercialization relationships with research universities, some of which may not be located locally. The development of creative talent from high-quality human capital programs can add comparative advantage to the assets of a region (Florida 2002a; 2002b; 2005a; 2005b; Florida and Gates June 2001). As the region and its localities establish and embed an "enterprise culture" (Refer to Part V, section on Relational Planning Concepts, A–Z, pp. 198–9.) to enable the community to compete more effectively in the global knowledge economy, science, technology, culture, creativity and innovation need to be promoted and celebrated routinely to demonstrate, across the generations, how critical these factors are to the long-term economic well-being and continued creativity of the community (Scott 2000).

Within the context of effective planning for amenities and quality-of-life factors, strong institutions in general, and responsive nongovernmental organizations, nonprofit institutions and voluntary organizations in particular, can be critical to the strategic positioning of cultural activities in a region's development (Beito *et al.* 2002). The nonprofit sectoral link is worth noting for a number of reasons: (1) nonprofits provide part of a community's cultural/creative environment, and offer an organizational form that allows creativity to be institutionalized when no other such outlet is possible; (2) nonprofits are employers and therefore can be important players within the urban economy; and (3) nonprofits, especially in the US, have been able to be more experimental in their work than government. Nongovernmental organizations frequently have piloted and executed trials of experimental policies.

The literature on nongovernmental institutions offers planners some findings that, with local tailoring, may be useful in framing sustainable amenities and quality of life initiatives. Issues of the arts, creativity, innovation and information technology in the US have been addressed by recent publications (Americans for the Arts 2003; Cohen *et al.* Spring 2003; Markusen Summer 2004; Markusen and King July 2003; Markusen *et al.* March 2004; Mitchell *et al.* 2003). Planners can draw on several recent resources that provide insight into the structure and dynamics of the nonprofit sector in a particular case region. The research conducted as part of the New York City Nonprofits Project has explicated the large, dynamic, growing, and economically significant nonprofit sector of the city (Seley and Wolpert May 2002). The book, *Measuring the Impact of the Nonprofit Sector*, provides an overview of the many methods and disciplinary approaches that can be used to study and understand the highly diverse nonprofit sector (Flynn and Hodgkinson 2001). In the US, cultural policy and the arts issues are researched and scholars are trained to study and work in the emerging field of cultural policy and the arts. The Center for Arts and Cultural Policy Studies provides such services. There is also a digital archive of data on the arts and cultural policy. Princeton University hosts both of these resources (Center for Arts and Cultural Policy Studies (n.d.); CPANDA (n.d.)).

The proximity of high-quality natural environmental attributes also represents assets that are attractive to knowledge workers and research and development organizations and firms. A place such as Seattle, for example, offers both sets of amenity factors – i.e. both social-cultural functions and natural environmental qualities (Sommers *et al.* December 2000; Sommers October 2, 2002). In addition to quality natural amenities, including low levels of pollution, Markusen, Hall and Glasmeier early identified empirically the following social, cultural, and institutional amenity factors for attracting high-technology personnel and enterprises: an airport with good connections for passengers and cargo; good housing conditions; good educational opportunities; specialized cultural services and recreational opportunities; regions that are weakly unionized, have low wage rates and high rates of unemployment; areas with good connectivity and transportation linkages; specialized business services; and "places with anti-regulatory, free enterprise ideology" (Markusen *et al.* 1986: 132–143).

The importance of amenities as an economic development tool at the regional level has been demonstrated for schools, the environment, crime and congestion (Gottlieb August 1994). Residential amenities need to be considered in such planning, both at locations near the workplace, and also where knowledge workers are likely to live (Gottlieb 1995).

Some of these amenity and quality of life factors demonstrate patterns of dispersed spatial organization; these include the routine nonprofit and governmental social services institutions that function in most communities. Other factors, such as specialized cultural institutions, as in the case of museums, are place-specific and often cluster in cultural districts or other concentrations in urban areas (Errington *et al.* 2001). Natural environmental amenity factors, similar to primary economic activities such as mining, take on spatial organizational patterns that are place-specific and therefore assume concentrated spatial organizational distributions. Some natural resources, such as forests, possess attractive attributes and therefore have strategic development potential.

There is a final note of caution relative to the e-Business Spectrum: from the initial experience of the dot.com bubble, it was learned by many investors the hard way that fundamental business practices cannot be disregarded just because information technology is being employed to do business. In the end, turning a profit is the goal of doing business. If electronically enabled business practices are not leading in the direction of profit or societal benefit, then a re-thinking of the approach or the business model is indicated.

Space and Time: Moving Theory into Planning Practice

The second of Graham and Healey's (October 1999) four guiding points for practicing relational planning is the need for planners and regional stakeholders to stress the multiple meanings of space and time. The production and consumption functions described above serve as the economic development drivers or the independent variables of the e-Business Spectrum. However, in order to be helpful and robust enough for the purposes of strategic guidance, intervention and planning, the e-Business Spectrum requires additional dimensions to be applied. These include, among other dimensions, relational spatial models and relational planning process models. This bundle of relations and processes in the form of e-business functions and factors, combined with spatial and planning process models, can be useful to local and regional planners as they engage in the framing of strategic planning with the new blend of global knowledge economy and network society relations.

One of the organizing principles adopted here in stimulating a new planning mindset and the application of local economic development theory is ubiquity – i.e. "no part of the planning region should be left behind in doing local planning in the context of today's global knowledge economy." The multifunctional and multilocational e-Business Spectrum therefore is an organizing framework that suggests explicitly that there is a wide range of ubiquitous knowledge economy development opportunities that is available for assisting in strategizing and implementing in most regions and localities. By blending the functions of the e-Business Spectrum with the pertinent location theories, local and regional planners may be able to formulate a variety of theory-based or intelligent strategies and tactics for effectively exploiting the knowledge economy development potential of their localities.

Planning practitioners can turn to a noteworthy body of local economic development location theory – i.e. theory that has been derived from reality and tested empirically. The models discussed here are explicitly spatial. The discussion draws heavily from the recent writings of Paul Plummer and Mike Taylor. Their work was selected because they have covered the principal location theories and models of external regional economic growth, and because they developed measurable dimensions that are intended to encompass the primary principles of the theories. In turn, they empirically tested the validity of the models and measures on regions in Australia (Plummer and Taylor 2001a; 2001b; 2003). This body of work can be helpful to planners in operationalizing Graham and Healey's guiding point on the multiple meanings of space and time.

Plummer and Taylor have examined six spatial theories of local and regional economic development. These include:

1 growth poles and growth centers (spillovers and spread);
2 product-cycle model (industry life-cycle);

3 flexible production and flexible specialization;
4 learning regions and innovative milieux;
5 competitive advantage;
6 enterprise segmentation and unequal power relations.

Refer to Part V, section entitled Relational Planning Concepts, A–Z, pp. 189–220. In that section, each of the models or theories is elaborated. We turn now to additional spatial relational ideas.

Space: Working Composite Spatial Construct of Local Economic Development

In order to simplify and make the many elements of these theories operational for the purpose of conceptualizing planning designs and applications, a working composite construct was formulated. It is outlined here. Across the above spatial theories, the firm is central to local economic development. In some cases it is existing firms that are critical to a theory; in other cases, new firms are the objects of the theory. Size of firms sometimes is determinative; some models are large-firm based, others are small and medium-size enterprise (SME)-based. In some cases, the firms and industries emphasize local markets; in other cases, the firms target external markets at the national level and globally.

Some theories are derived from urban areas and others explain economic growth in areas of new development and by means of new industries. Generally, however, these theories have the clustering of various economic activities as the primary stimulus for growth.

To the composite, the competitive advantage model contributes and draws attention to, and notes the importance of the role of the organization. Organizational factors include: capacity; flexibility; motivation; strategy; management; commitment; and competition. Networking plays a role in some of the theories. Several of the theories make explicit the role of time, timing, and the rate and pace of change processes in local economic development.

Derived from the six local economic development theories, some of the principal factors of the working composite spatial construct include:

- Human resources and their local specialization; skilled, stable jobs, employment and continuous training.
- Technology leadership and its local control.
- Access to information and knowledge resources for creation and knowledge transfer.
- Capital resources and investment.
- Institutional support.
- Other infrastructure – e.g. physical and social.

- Organization and organizational behavioral factors as noted above.
- Network relationships that support the locational integration of large, economy-driving, high-impact firms and supporting small and medium-sized enterprises (SMEs) – i.e. as suppliers or innovators.
- Market orientation: local market, or interregional trade and global market, or both local and global.
- Timing, speed, early-stage innovation, rapid, pace, planned change.
- Spatial organization and locational pattern, cluster and concentration, or dispersion, or both, and uneven or even distribution.
- Business power and competitive relations across regions and localities.

Space: Theory Integration, Benchmarking and Measurement

In the review articles cited above – i.e. of the six local economic development conceptualizations – Plummer and Taylor have offered a treasure trove of theory and models to stimulate strategic and tactical ideas for practicing planners to ponder and consider engaging for tailored application to their own region and localities. While the location theories noted above are not exhaustive, they are representative of most of the tried and tested contemporary and effective approaches to technology-based spatial economic development at the substate scale.

These contemporary development theories offer planners the promise that regional growth is a result of policy planning and implementation that is intended to influence technology-based development conditions within the region. Such planning requires detailed and specific knowledge of the region's assets and development potential. Thus, issues of measurement and policy evaluation also must come into play.

Plummer and Taylor concluded that of the six theories that they reviewed and assessed, "each involves different permutations and combinations of eight dimensions that currently are thought to enhance local economic capacities to create growth and to cope with change" (Plummer and Taylor 2001a: 228). The eight empirical dimensions that Plummer and Taylor derived are listed below. Also included below is the associated available surrogate measure and year that they were employed from Australian data sources for each dimension:

1. Technological leadership at the enterprise level:
   measure: index of high-technology industries (data from late 1980s).
2. Knowledge creation and access to information:
   measure: index of access to information (data from late 1980s).
3. Local integration of small firms:
   measure: percentage of establishments in multilocational enterprises (from 1992).

4 Infrastructure support and institutional thickness:
   measure: effective protection rate (in 1990).
5 Local human resource base:
   measure: percentage of working population without a degree (in 1991).
6 Power of large corporations affecting structure and strategy:
   measure: index of corporate control (from 1992).
7 Interregional trade and the extent and nature of local demand:
   measure: index of intermediate market accessibility (data from the late 1980s).
8 Local sectoral specialization:
   measure: index of specialization (from 1990) (Plummer and Taylor 2001a: 228).

Plummer and Taylor concluded that no one dimension by itself produces local economic growth. Thus, they reasoned further, there is no one path to successful local economic growth in the global economy. As a result of their empirical modeling exercise, they identified two sets of processes that served as the basis for local economic growth in Australia:

- The magnitude of local human resources (or human capital development).
- The local presence of an enterprise culture, built principally on technological leadership but with an element of local enterprise integration.

They concluded that these two sets of processes operate separately and "they can also be looked on as processes at two discrete stages in the value chain that creates new knowledge in a city, region or community" (Plummer and Taylor 2003: 644).

Such measures are directly relevant to substate regional planning today. In the United States for example, given the strong emphasis of the US Economic Development Administration (EDA) for results oriented regional planning, use of economic development measures is critical to the award of grants and support for local regions or Economic Development Districts (EDDs). Thus, a major operational and strategic challenge facing regional planners in each state of the United States is the identification of readily available development measures and/or their available surrogates. Measures then must form the basis for evaluating local plans according to the EDA's Investment Policy Guidelines – i.e. they state that proposed investments should be: (1) market-based; (2) proactive in nature and scope; (3) look beyond the immediate economic horizon, anticipate economic changes, and diversify the local and regional economy; (4) maximize the attraction of private-sector investment that would not otherwise come to fruition absent EDA's investment; (5) have a high probability of success; (6) result in an environment where higher-skill, higher-wage jobs are created; and (7) maximize return on taxpayer investment (Sampson April 2, 2002).

Benchmarking

In such a policy context, an important purpose of benchmarking is to improve the region's planning by learning from others. It involves regular and routine systematic comparison of planning performance with the best such planning and planning organizations elsewhere. Benchmarking enables the identification of gaps in performance; these findings can be used continuously to improve the process and outcomes of the local target planning. Such behavior can lead to the development of learning organizations and learning regions. In Michigan, for example, seven "Competitiveness Vectors" have been identified. These measurements are intended to be used diagnostically to improve the state's overall technology-led economic development performance, competitiveness, and therefore its development: (1) human investment; (2) financial resources; (3) innovation resources; (4) infrastructure; (5) business costs; (6) globalization and vitality; and (7) quality of life (SRI International and Michigan Economic Development May 2002). These measures may be compared to the functions of the e-Business Spectrum as well as to the measures employed by Plummer and Taylor above, thereby enabling the practicing planner to see the subject-matter overlaps and gaps. These and other comparisons can facilitate the planner in the construction of a relevant measurement system for her or his own local region. For a useful comparative resource on "the world's most effective polices for the e-economy," see Booz Allen Hamilton November 19, 2002.

Measurement

In trying to understand the new dynamics of city-regions of all sizes around the world, Jean Gottmann, in 1974, observed that occupational statistics and shifts in the structure of employment need to be analyzed. He went on to suggest that the old classifications of primary, secondary and tertiary economic activities should be reformed to break out the work that is engaged with the manipulation of information that leads to, or involves the making of abstract transactions. He called this "quaternary" work. He wrote "the size and role of the quaternary personnel is best evaluated on the basis of occupational rather than industrial classification" (Gottmann 1974: 257). The emphasis on occupations as well as industries has been continued and extended by Ann Markusen. She has targeted performing arts occupations, with both quantitative and qualitative occupations evidence to both advance location and development theory. She specifies how planners too may target occupations in their practice (Markusen Summer 2004).

In addition to occupational data, regional and local planners today may use standard government industries data series (e.g. North American Industrial

Classification System (NAICS) and, historically, Standard Industrial Classification (SIC)) to go beyond the traditional concepts of primary, secondary and tertiary industries, and regional economic models with their traditional focus on the role of manufacturing. Charles Colgan has demonstrated this new approach in measuring the "creative economy" and the "ocean economy" components of the regional economy (Colgan October 2004). From his applications of this approach to the US state of Maine, Colgan has observed that the local planner faces three principal constraints: "Inherent limitations in the data sets themselves and the taxonomies used to organize them, the shifting taxonomy of SIC to NAICS, and the strict rules governing the release of confidential data" (Colgan October 2004: 27). Despite these constraints, he concludes that the new ways of conceptualizing regional economies, as in the case of cluster theory, and with new readily available and relatively inexpensive technologies, such as geographical information systems and the processing of geo-coded local data, combine to empower regional and local planners to define, measure and track selected components of the new economies at the subnational, substate levels and postal-code levels.

In the United Kingdom, an IT trade association, Intellect, has stated that "The UK is using measurements based on outdated business/economic models, which fail to assess our progress towards the new knowledge based economy and threaten to undermine policy planning" (Intellect n.d.). As a consequence, Intellect, along with the Institute of Directors, the London School of Economics, the BBC, Intel, Apple, BT, Hewlett-Packard, and others, has created an "Intellect Index" that measures progress towards the trade association's vision for a knowledge-driven economy. In developing the index and by conducting gap analyses, this partnership seeks to improve the government's measures and data collection to ensure congruence between policy goals and results. Such independent auditing is a best practice that should be considered by the opinion-setters of all regions.

For other examples of local planners and business practitioners who are trying to meet the challenges of theory-inspired measurement and benchmarking, see the efforts by Smart Michigan (n.d.) and Mahendhiran Nair (2004). For planning regions in Michigan, Rex LaMore and the Knowledge Economy Research Team have developed regional and local knowledge economy snapshots for assisting planners and their stakeholders to become aware of some of the implications for their respective city-regions (Gandhi *et al.* June 2005; LaMore *et al.* July 2004; April 21, 2005). Business professor Mahendhiran Nair has modeled socio-economic factors for use in measuring competitiveness in the context of developing countries (Nair 2004).

Time: Relational Program Planning

As part of the multiple meanings of time, a non-linear staged relational planned change process construct is introduced here as an illustration of a time relationality in operationalizing the Graham and Healey organizing framework and for its usefulness for practicing planners. In order to affect the planned changes required to realize the social and material advancement potential that is embedded in the e-Business Spectrum framework, ideally, a relational planned change process is required. The Program Planning Model (PPM) is such a relational planning process; it has proven to be effective (Van de Ven and Koenig, Jr Spring-Summer 1976; cf. Bryson 1988). PPM is appropriate in this context for several reasons: it is congruent with the body of relational theory that is used throughout this book; also, it has been applied and tested in various institutional and organizational contexts. This planning process is distinctive in that each of its phases or stages is led by particular actors who perform specific roles that are unique to, and/or dominant in each planning phase. The outcomes of each phase contribute to make the whole planning process more value-added than the sum of its parts. Refer to Part V, section entitled "A Time-Relational Method: The Program Planning Model," pp. 237–9. The strategies that emerge from effectively implemented PPM planning behavior can have a high impact and be transformative. These behaviors, in the words of Patsy Healey, are "not linear step-by-step processes. Episodes of transformative strategic potential rise up when moments of opportunity appear" (Healey July 2005: 26). The notion of "moment" is elaborated below in the next section.

One of PPM's principal values is its effectiveness in breaking down complex planned change and policy processes into understandable and actionable stages by which specialized actors may perform specialized roles within a coordinated planner and stakeholder team context. The Program Planning Model offers planners and managers an effective method for introducing innovation and for steering intentional planned change in dynamic, uncertain and temporary societal, community and organizational environments. These conditions are the essence of time relationalities. PPM consists of seven principal stages; each of these has lead actors who perform specific roles that relate, complement and contribute to the other roles and actors of the planning process. Feedback and learning from stage to stage is an operating principle in PPM; focused piloting or controlled trials can lead to broader experimental demonstration, and these results can form the basis for full program implementation. PPM also has an organizational management complement for innovation – i.e. the Program Management (PM) perspective (Delbecq and Van de Ven March 1971).

Time: The Moment is Important in Planning Practice

As part of operationalizing and practicing multiple meanings of time, the concept of the "moment" has emerged as an important organizing concept for relational planning. Within the context of developing a process-based methodological framework for a research-driven new economic geography, Henry Yeung has noted that there are "different 'moments' of a relational research process that are sensitive to specific research questions and/or contexts" (Yeung 2003: 442). He has elaborated:

> The word 'moment' is used intentionally for two reasons. First, it demonstrates my commitment to a nonlinear and nonsequential way of under-

> standing the research process. In my framework, any moment can be important and prioritized by a particular researcher, depending on the kinds of research questions asked, the theoretical approaches taken, and the research context. Second, and in line with my implicit acceptance of new economic geographies, I want to keep my methodological framework open and dynamic, while also recognizing the complex interrelationships among its constituents.
>
> (Yeung 2003: 458)

Planning researcher Patsy Healey has observed that the regional planners of Milan, Italy, have adopted the practice of waiting for the moment as they seek to implement their Framework Document (Healey September 111, 2003c). In the complex environments of the network society, as these dynamics play out on the ground, opportunities and developments may evolve and can emerge gradually. So, an approach that may prove effective for planners today is to take the long-term perspective, be sensitive to timing, and thereby be positioned to be able to seize the moment of opportunity as and when it might emerge. In the case of policies for the development of cluster advantages for example, Maskell and Kebir have stated that such "results are measured in decades if measurable at all" (Maskell and Kebir 2005: 13).

Time: Path Dependence

The path-dependence concept is useful in reminding us that the development history of a place and region is quite likely to be significant when planning for future development. Past technology and economic decisions and advantages that emerged as a result of those decisions, for example, may serve as a basis for continued niche development strategies. Future development may depend on, or be highly influenced by such decisions from the past, or future strategies that break away from such dependence may be suggested. Legacy planning too often is a barrier to change and has operated to produce a planning mindset that needs to be unfrozen before new and different approaches to planning may be accepted and put into operation. These issues are compounded by the normal difficulty of attempting an approach to planning that is new and formative. Refer to Path Dependence in Part V, section on Relational Planning Concepts A–Z, pp. 211–12.

Time: Evolution and Maturation

Time relationality in development also has important longitudinal characteristics. For example, one may track the progression of ICT-facilitated urban and regional development in Southeast Asia by analyzing planning strategies and their outcomes over the last generation. Mature strategies may be compared to more recent plans. Such analyses permit interpretations of leapfrogging technology-development stages, or the diffusion and replication of technology-based programs, or changes in the pace and direction of ICT access and roll-out; or shifts in priority from the provision of digital infrastructure to the development of creative content that is enabled by such infrastructure (Corey 1998; Corey and Wilson 2005b).

Timelessness

Another dimension of time relationality are concepts that are not necessarily restricted to a particular time. Jean Gottmann, for example, used concepts that were pertinent whether they were applied to ancient times or to the most modern periods of urban and regional development. Among others, these concepts included: hinge, crossroads, megalopolis, and so on. Refer to Part V, see section on Gottmann Concepts, A–Z, pp. 221–7.

Future Time: Global Planning and Futures Context

Increasingly, it is the global context and scale within which regional and local planning will be conducted explicitly in future. Are there planners who have thought and operated at such a scale and for the long-range future? From a global and visionary perspective, the late Greek architect-planner Constantinos Doxiadis may be inspirational to us collectively in stimulating creative planning-practice conceptualizations to ensure that a long-term futures dimension, as well as a global dimension is incorporated into city-region scale strategic planning. Doxiadis invented and operationalized the highly integrative and synoptic field of Ekistics. His Ekistics grid depicts a global to local to individual-person layers, similar to those discussed throughout this book. He is one of the few modern planners who has conceived and worked at the global scale as an overt part of his planning-practice mindset (Bromley July/August 2002; September/October 2002; November/December 2002; 2003). He and Jean Gottmann were colleagues who interacted regularly at the Delos Symposia that were organized by Doxiadis. His concept of the global-wide city, or ecumenopolis– i.e. "the coming city that will cover the entire earth" – offers a vision by a planner for planning that is global in scale and futures in time perspective (Doxiadis 1966: 87; 1968; 1977). In terms of the world-wide digital development infrastructure network represented by the Internet, the global knowledge economy and network society already today has permeated the business and living environments of much of the planet, thereby leveling the competitiveness playing-field (Friedman 2005a). This pervasiveness should serve as a wake-up call for all planners and their client stakeholders to become aware of the new development implications from ICT-stimulated changes and to take informed and planned action to execute appropriate intelligent development strategies for the futures of their respective city-regions and localities (e.g. Doxiadis 1967; 1969; 1970).

Multiple Layers: A Spatial Relational Planning Framework Hierarchy

Graham and Healey's third guide point for the practice of relational planning indicates the need to "*represent places as multiple layers of relational assets and resources, which generate a distinctive power geometry of places*," and "*recognize how the relations within and between the layers of the power geometries of place are actively negotiated by the power of agency*" (Graham and Healey October 1999: 643, emphasis added).

Relational layers of power geometries may take several forms. They may be thematic as in the case of levels of educational institutions and degrees of community economic distress and equity. Layers also may be temporal, as in the case of past development issues that vary over multiple time periods. Layers further may be conceptualized as spatial distributions of development content at various scales. For example, a range of knowledge-economy topics may be ordered by a spatial hierarchy from macro-scale to micro-scale. Such topics as governmental policies or settlement forms may be organized from global-scale urbanized settlement patterns down the hierarchy to national, to sub-national, to local scales of urban settlements (Doxiadis and Papaioannou 1974; Doxiadis 1968). Examples of such spatial-distributional layers are illustrated below in Part IV, see pp. 114–26.

For the purposes of effective planning practice in the global knowledge economy and network society, the spatial-locational unit for targeting is the city-region (Scott 2001). It is the functional linkages and interdependencies between the city and its region that is a principal layer for policy and strategic planning attention (Greenstein and Wiewel 2000; Pastor, Jr *et al.* 2000). In this era of high levels of electronic connectivities, these city-regions increasingly are networked and form a new spatial layer that spans the global economy. These inter-city-region relations form this digital development and intelligent development layer, especially by means of the diverse flows of financial and business services among global and regional firms across the world (Taylor 2004a). In the earlier phases of the evolution of the global knowledge economy and network society, it was the largest city-regions – i.e. world cities and global cities – that were researched and better understood (Hall 1997; Sassen 1991; 1994). The world city-region of London is illustrative of the kinds of extensive research and study that can be conducted to provide planners and their stakeholders with an in-depth understanding of urban and regional dynamics of city-regions, including those of world city and global city influence, as well as those of more limited scale (cf. Buck *et al.* 2002; Ellis *et al.* 2002; Hamnett 2003; Taylor April 15, 2004b; for background, also refer to Ackroyd 2001). Thus, no longer are inter-city hubs and flows of the global knowledge economy the sole domain of world and global level city-regions. This is a function of the world city-regions, with their highest economic functions and multiple roles, as well as the global cities as centers of financial and business services, and production activities dispersing some of these functions, especially the more routine functions to locations that occupy lower layers in the city-regions hierarchy. These dynamics are a function of the forces of globalization and informatization – i.e. the increasing pervasiveness of ICTs to most economies and societies, large and small. Peter Taylor and his colleagues of the Globalization and World Cities Study Group and Network

(GaWC) have been researching these hierarchical tendencies and networking relationalities among the world's city-regions (Taylor 1997; 2001; 2004a). The GaWC's *Atlas of Hinterlands* is noteworthy for its clear depictions of the networks among the world's city-regions (Taylor n.d.)

Over the last several decades, as many knowledge-economy functions have migrated down the urban hierarchy from global-level activities to more regional and local-level activities (Markusen *et al.* 1999), planners and leaders of lower level, smaller settlements and environments now may consider developing planning-support systems for government and business functions, and services that heretofore were economically viable principally at higher-level settlement forms. For example, in the United Kingdom (UK), there is a policies hierarchy for knowledge-economy policy initiatives and environments that can stimulate innovations and analogous strategic planning elsewhere. Compare the Intellect Index (Intellect n.d.) to programs of the UK's National Endowment for Science, Technology and the Arts (n.d.), to the London Development Agency's (n.d.) economic development strategies and initiatives, and to the policies and programs of the UK's Countryside Agency (n.d.), and to Taylor 1999. Today, then, some of the planning approaches and tactics that heretofore were considered applicable exclusively to global-level city-regions – e.g. London – have wider applicability to places and settlement layers that are much lower in the spatial hierarchy. For example, leaders of all city-regions seeking to enhance their competitiveness in the global knowledge economy should have effective personnel – e.g. an e-envoy – assigned to ensure that plans for digital development and intelligent development priorities are being implemented and evaluated. Smaller regions also may benefit from maintaining a function devoted explicitly to attracting and retaining jobs and investment, as in the case of the mayor's London Development Agency.

All major types of regional environments need such digital development and intelligent development planning and plan-implementation attention. Other city-region ecologies or layers that require explicit global knowledge-economy strategic attention include: metropolitan regions other than the largest global city-regions (Ozawa 2004); downtowns and city centers (Florida 2005a); neighborhoods (Hampton 2003); suburbs; edge cities (Garreau 1991); edgeless cities (Lang 2003); small towns (Cohill and Kavanaugh 1997), and rural regions (Munnich, Jr *et al.* 2002). These relational layers are relatively straightforward and static; Stephen Graham has called attention to additional, more complex dynamics that he has labeled "flow city." He has identified four "networked mobility spaces" that have emerged in today's metropolitan regions: "e-commerce spaces, passenger airports and fast-rail stations, export processing zones, and multimodal logistics enclaves dedicated to freight" (Graham 2002: 3). He utilized these spaces to demonstrate that increasingly, the metropolis today hosts enclaves that have locations in the city-region, but they are disconnected from it and do not function of it as much as they are connected to other such enclaves located elsewhere. These kinds of relationalities are illustrative of some of the new locational and spatial dynamics that are associated with globalization. These more complicated layers are yet another local reflection of the global knowledge economy and network society.

Power of Agency: Transforming Digital Development into Intelligent Development

The fourth and final guiding point of Graham and Healey's call for practicing planning in a relational way is to "recognize how the relations within and between the layers of power geometries of place are actively negotiated by the power of agency through communication and interpretation" (Graham and Healey 1999: 643).

For an operational usage of "agency," refer to Part V, section on Relational Planning Concepts A–Z, p. 189. This guiding point may be illustrated by drawing on our experience in Michigan. In order for a regional planning organization in the state to be awarded planning and program implementation funds under US Department of Commerce Economic Development Administration (EDA) programs, the "agency" for these planners is the mandated Comprehensive Economic Development Strategy (CEDS) document. These strategic planning documents must be prepared and kept updated in order to have an individual planning region or Economic Development District (EDD) be eligible for EDA funding. The CEDS document therefore represents an immediate and tangible structure of agency within which practicing regional planners in the United States can ensure that their planning is proactive and responsive to the demands and complexities of the global knowledge economy and network society. The revision and updating of a region's CEDS should incorporate access to, and full exploitation of ICTs throughout the region and its localities. Planning for both ICT infrastructure and the more conventional physical infrastructure are fundamental to a "knowledge economy based CEDS." With such contemporary infrastructure systems in place and being continuously updated, planners may have positioned their region to be more competitive in the global networked economy and society. The planning can focus on e-business content and its creative application to stimulate *development*, but not only economic development. The planned development of the region should strive for the incorporation of values relations within its development – e.g. reduce disparities and seek equality.

Power of Agency: Development – The Values Imperative

The ultimate impetus for this planning practice work is development – i.e. multifunctional, holistic, equitable and relational development. Everett Rogers has offered an operational definition of such development that fits the purposes of this discussion – i.e. it provides the needed values imperative. He defined development as:

> A widely participatory process of social change in a society intended to bring about both social and material advancement (including *greater equality*, freedom, and other valued qualities) for the majority of the people through their gaining greater control over their environment.
> (Rogers 1976: 225, italics added)

This statement captures the essence of the policy intent of the US Economic Development Administration regarding "economically distressed areas" and the measures required to address such distress. The statement also encapsulates the principal functions, factors and relationships of the e-Business Spectrum conceptual framework.

Power of Agency: Mediating Power Layers and Planning Scenarios

As can be demonstrated via mapping, the distribution of the knowledge economy assets and resources of a state and its regions are not evenly distributed – i.e. they represent multiple power geographies. For example, such functional power layers as the following might be mapped, from the most highly specialized to the more numerous assets: locations of the highest ranked knowledge economy regions (Huggins *et al.* 2004); locations of biotechnology and pharmaceutical firms; locations of smart zones technology clusters; locations of biosciences firms; locations of tertiary education institutions; and the location of empty areas and areas of sparse development. For these layers, there is a great deal of functional and spatial unevenness across the regions and localities of any state or province. The unevenness of various spatial relational knowledge economy layers may be mapped and analyzed, along with other empirical trends and their trajectories. In order to address the EDA's distressed communities requirement, for example, and to illustrate with the Michigan case, regional planners should incorporate two additional mediating layers – i.e. (1) the locational pattern of the state's e-responsiveness regions (e-responsiveness is a regionalization of the state's areas of e-readiness and quality of the areas economic development websites); see Breuckman April 2003 and Singh July 2003, and (2) the locational pattern of the state's economically distressed areas. These two layers may be used to mediate and set priorities for planned scenarios and policy interventions, such as targeted investment and action programming by public, private, nonprofit and individual regional stakeholders, by means of local partnerships and alliances. From mediating the functional power layers by means of the two mediations of economic distress and degree of e-responsiveness, four planning-scenario outcomes may be derived for priority setting and targeting for those localities that are: (1) most e-responsive and economically distressed; (2) somewhat e-responsive and economically distressed; (3) least e-responsive and economically distressed; and (4) most e-responsive and not economically distressed. Refer to Figure 17 on p. 118. Such priorities and their sequencing can vary from this scenario, but the fourth Graham and Healey relational planning guiding point can be demonstrated to be operational by such strategizing.

Old and New Mindset

Some Lost Traditions of Planning

One of the finest traditions from the history of city planning was to ensure that plans have a long-range dimension. Two generations ago, it was not unusual for some of the long-term planning periods of city plans to extend beyond 25 years. Typically, such plans had a utopian, idealistic and futuristic theme – e.g. Reiner 1963. An inspirational example, that brought together these themes, was Percival Goodman and Paul Goodman's 1947 book, *Communitas*. The Goodmans formulated three regional-scale future-planning alternative and comparative paradigms or models, or scenarios of community life. This is the kind of planning scenario formulation that made these early planning thinkers and practitioners role models for inventing creative and pragmatic ways for community actors to better harness the potential of the unknown future.

During the last generation, however, both the long-range futures and the utopian dimensions have waned to the point of being rare to nonexistent in mainstream planning and in the plans themselves. Today's plans and planning have become short-term and pragmatic, often with minimal vision. Urban and regional plans today have evolved generally to be composed around short-term horizons. Indeed, today's planning time periods often resemble annual and other near-term plan and budgeting periods – i.e. two to five years.

Given the complexities of today's economic and high-technology dynamics, and given the long gestation periods associated with such instrumental functions as the research and development of some products before they can reach the market – e.g. pharmaceuticals – a case can be made for much longer time horizons for contemporary strategic planning. Given the chronic disparities and inequalities in and among many countries and their cities and regions, today's planning and plans also would benefit from explicit idealistic and human development priorities. These are not either/or issues. Today's regional and local planning requires pragmatism and idealism, as well as short-term and long-range plan making. For example, human and intellectual capital development, including life-long learning, requires sustained generation-long public and private investment. People going through early childhood education and on to advanced graduate higher education require the time to complete this human development cycle and to sustain such growth throughout the remainder of their increasingly longer lives. The reintroduction of the long-term view and idealistic values in the context of future planning are ingredients that should enhance the plan-making of today's planners and their stakeholders in their objectives of facilitating their region to be more competitive.

New Planning and Old Planning

We are fortunate today to have a wealth of new ideas and approaches from the scholars and planners who have contributed to the body of concepts discussed in this book. It is worth noting that these recently developed planning concepts are not the only ones that can be useful in meeting the challenges of today's relational planning requirements.

While not explicitly labeled as such, relational thinking was practiced by a number of intellectually influential individuals who might be considered to be representative of some of the best relational mindsets and practitioners of earlier eras of city planning, and the study of cities and urban regions, and their organizations. These earlier contributions have been helpful here in enabling us to construct a useful model for enhancing the practice of urban and regional planning in the global knowledge economy and network society. The concepts of Jean Gottmann have been particularly valuable. Other earlier relational thinkers and practitioners who have also influenced our work include: Chris Argyris (e.g. actionable knowledge); Constantinos Doxiadis (e.g. Ekistics); Paul and Percival Goodman (e.g. idealistic scenarios); Donald Schön (e.g. reflective practice) and Melvin Webber, among others. For exposure to some of these early relational thinkers and their concepts, refer to the Reference section and Part V, especially the section on Relational Planning Concepts, A–Z – for example, see Webber's concepts of "localite" and "cosmopolite," pp. 209–10 and p. 196.

These and many other earlier contributors are owed our collective debt of gratitude for their prescient work. By means of their publications, their influence lives on and continues to contribute to our intellectual growth and knowledge. For example, more than ten years after his death, the contributions of Jean Gottmann enable us to better engage the new planning challenges of this new age. Only last year, an international conference on Gottmann's thinking was held in Paris. The theme of the meeting was entitled "Dedicated to the Geographic Thought of Jean Gottmann in Relation to Current Political, Cultural and Urban Challenges." Preceding the meeting, the periodical *Ekistics: The Problems and Science of Human Settlements* published three special double issues under the theme of "In the Steps of Jean Gottmann." Calogero Muscara was the guest editor for these issues (Muscara Jan./Feb.–March/April 2003a; May/June–July/August 2003b; Sept./Oct.–Nov./Dec. 2003c).

For the stimulation of our planning thinking, Patsy Healey represents the new planning contributions. For several decades, her analyses and interpretations of regional communicative and governance issues brought her, along with colleague Stephen Graham, to the point of being positioned to construct a much needed model for relational planning practice (Graham and Healey 1999). Healey has continued to observe and interpret city-regional developmental case studies in Europe (Healey July 2005). This scholarship has enabled her to provide valuable empirical insights for assisting us collectively to begin to understand the dynamics at the interface of regional planning and development in the global knowledge economy and network society (Healey forthcoming).

Inspired by the earlier concepts of Jean Gottmann and some of his generation of urban planners and urban studies researchers and thinkers, and inspired by the new concepts and conclusions of Patsy Healey and some of her generation of planners and scholars, we are all better prepared to engage the new unknown relational future. Both the old and the new, in creative combination and tension, will inform the development of our approach to city-region scale planning in the global knowledge economy era.

Part III
Context

Introduction

Many around the world express surprise that the country in the world with the highest level of broadband penetration is South Korea. The fact is that this kind of science and technology advance was generated by recent South Korean leadership, innovation, and investment priorities; and the early results were surprising to many. While such technology and science-driven policies in South Korea were adding value and new opportunities to its economy and society, many regions of the United States, for example, and many areas of other parts of the global economy – in both developed and developing areas – remain unserved, or underserved by affordable broadband access. Also, the institutions of many of these same areas are relatively inactive in stem-cell research because of values debates and related regulatory prohibitions. This is the kind of digital development and intelligent development unevenness that characterizes today's global knowledge economy and network society. This is the context; this reality presents opportunity and development potential for the regions and localities of the world with the knowledge and organized will to plan and to act.

From our observations, we have derived contemporary lessons about knowledge-economy development planning from wherever these explorations have taken us. As we have studied and reflected on our diversity of findings, we discovered that our searches have led us around the world. Our quest for those ingredients to use in the improvement of the practice of urban and regional planning within the context of the global knowledge economy and network society has taught us that effective development planning practice can originate anywhere. Consequently, for continuous career-long and life-long learning and development, we strongly encourage local practicing planners to be open to receive innovative planning ideas from anywhere in the world. In turn, one should also be open to share one's innovative planning approaches with others, especially those outside your immediate local sphere of influence; thereby, experimental learning and improved new planning practice may be tested and advanced. In the process, the beginnings of the formation of a global planning-practice learning community may be put in motion. Myers and Banerjee have observed:

> The context of planning – the world around us – is changing rapidly and dramatically with globalization, information technology revolution, the emergence of a new economic order, and the new prevalence of cooperation between government entities and among public, private, and nonprofit sectors.
>
> (Myers and Banerjee Spring 2005: 128)

Since the subject of this book is regional and local planning practice for technology-enabled development and its extension into innovation-inspired planning – i.e. intelligent development – we have relied principally on the developed regions and places of the world for context and comparative perspectives. For these purposes, the three main technology-economic regions of the global economy have served as our primary locational contexts of inspiration. These regions include North America, Eastern Asia and Western Europe. While this, Part III of the book, includes discussion and cases from each of these three regions, the other parts of the book also include material from these technology-economic areas. Part V and the References section of the book include additional commentary on places across the globe.

Context: The Three Global Technology-economic Regions

Globally, as measured by Huggins *et al.* (2004), the regions of the United States combine to make it *the* competitive knowledge economy of the world. Forty US regions make up the top 50 knowledge economy region-level rankings throughout the world. Among these 50, the only non-US regions, with their global ranking listed here in parentheses, are: Stockholm, Sweden (15); Uusimaa, Finland – i.e. the Helsinki region (19); Ile de France – i.e. the Paris region (34); Tokyo (38) and its neighbor region, Shiga (39), both in Japan; the South East of England, United Kingdom (40); West, Sweden (44); Switzerland (45); London, UK (46) and Eastern, United Kingdom (50) (Huggins *et al.* 2004: 6). The most competitive regional knowledge economy in the world is San Francisco (1); the Boston region is ranked next (2); followed by the knowledge economy regions of Grand Rapids–Michigan–Holland, Michigan (3); and Seattle, Washington (4). The Detroit–Ann Arbor–Flint, Michigan, region is ranked twelfth.

> The key factor underlying the selection of regions for benchmarking is their relative gross domestic product (GDP) per capita. In the main, the regions included are those that have achieved the highest output per capita across the globe during the recent period.
>
> (Huggins *et al.* 2004: 2)

The methods used by Huggins and associates generated a world-knowledge competitive index from 19 variables. One overall number – i.e. a composite index – was used to rank 125 city-regions or metropolitan areas of varying definitions from around the globe into the "World Knowledge Competitive Index 2004." Of these regions, 55 are in North America, 45 are in Europe, and 25 are in the Asia-Pacific region. The methods that produced the index are based on the use of factor analysis that permits the identification of an underlying structure among the set of selected variables. These variables were classified by five components: (1) human capital components; (2) financial capital components; (3) knowledge capital components; (4) regional economy outputs; and (5) knowledge sustainability (Huggins *et al.* 2004: 2–3). As published in 2004, Huggins and his associates ranked 125 largely city-regions using this index.

North America

North America is rich and deep in lessons, best practices and benchmarks for informing the needed new-practice of regional and urban planning in the new context of the global knowledge economy and network society. For our purposes, Canada and the United States comprise our operational definition of North America. For digital development discussion, Mexico is included in "Latin America" (Economist Intelligence Unit 2005). We selectively reference here some of the context faced by Canada and the United States, with full recognition of the limitations of this approach. Our goal here merely is to suggest the diversity and similarity of planning responses by policy-makers and business actors, among many other local stakeholders from the major technology-economic regions of the global knowledge economy and network society.

In the spirit of being suggestive, North America is the home to such digital-development and knowledge-economy pioneers, cases and even "models" such as: Silicon Valley, California, and its host state, California (Saxenian 1994; Hanak and Baldassare 2005; Zhang and Patel 2005); the e-city case of Virtual Charlottetown, Prince Edward Island (Virtual Charlottetown n.d.); the wired suburb of Netville, Ontario, of suburban Toronto (Hampton 2003); the arts and culture economic-development success story of Stratford, Ontario (Reid and Morrison 1994); or, among many other examples, the town of Blacksburg, Virginia, the first electronic village of the United States (Cohill and Kavanaugh 1997). Increasingly, as a relatively wealthy and technologically innovative and advanced couple of unique political-economies, the United States and Canada should be monitored by planners and their stakeholders so that development lessons are used when they formulate their own local planning processes and strategies.

Canada

Using the World Knowledge Competitive Index 2004, Canadian city-regions and their respective provinces are ranked as follows: Ontario (63); Quebec

(78); Alberta (80); Manitoba (92); British Columbia (96); and Saskatchewan (100) (Huggins *et al.* 2004: 6).

Innovation Systems

David A. Wolfe and his colleagues of the Innovation Systems Research Network (ISRN) and the Ontario Network on the Regional Innovation System (ONRIS) have analyzed many of the city-regions of Canada. Their body of work is noteworthy for our purposes here, because they have perfected a framework that is relational at its core and their findings may be translated by practicing planners and policy-makers into regional and local-scale innovative development programs. Importantly, their work is accessible to the practicing planner, local economic developer and regional stakeholder. For an example of the ONRIS Newsletter and its helpful links, see Ontario Network on the Regional Innovation System (2005). Wolfe and ISRN and ONRIS associates also have published books and many papers on the regional innovation system (RIS) framework and clusters. This work can inform practicing regional and local planners to see parallels with their own settings and to consider using analogous approaches in preparing themselves to engage the global knowledge economy and network society, and to tailor such work to change the local planning mindset. For representative books on this body of work, see Holbrook and Wolfe 2000; 2002; Wolfe 2003; Wolfe and Lucas 2004.

This work has been conducted within the context of the "innovation systems" conceptual framework. This framework is congruent with the study of the complexities and multiple relations of the knowledge economy. It is applicable especially as a means of understanding the innovation-production end of the e-Business Spectrum. The means of such production include new ideas, new goods, new services and new practices; in this context, the object ultimately is commercial success. It involves technology-facilitated information flows among people, enterprises and institutions. The innovation system framework can be applied to various sectors and scales from the national level (NIS) (Organisation for Economic Co-operation and Development 1997) to the regional (RIS) and local levels (LIS). Cluster development and the benefits of proximity are central to the innovation systems process. Refer to these concepts in Part V, section on Relational Planning Concepts, A–Z, pp. 189–220; see Clusters; Innovation; Innovation Systems; Ls – the Five Ls; Local Innovation Systems; and Regional Innovation Systems.

The application of the innovation systems framework within the context of the Canadian political economy reveals the power of the framework for contributing to understanding the many relations across sectors and at different layers of spatial organization. In order to enhance local competitiveness, innovation can and should be applied continuously to all appropriate sectors, both new economy sectors and old economy sectors. The innovation systems framework enables and accommodates cross-sectoral and spatial comparisons.

Canada as a technology-economic region is a useful case for our purposes here. The Canadian economy offers a wide variety of city-regions and localities. Some are remote and distant from large urban centers. For example, the province of New Brunswick is a lagging region. Traditionally, this province has relied on natural resources for its primary economic development. Davis and Schaefer (2003) have researched the initiatives and policies of the province to diversify its economy by seeking to establish an information and communication technology (ICT) sector cluster. "Cluster," as used in this context, is defined as "an aggregation of related firms in a geographically bounded area" (Davis and Schaefer 2003: 122). Also, refer to Part V, section on Relational Planning Concepts, A–Z; see Clusters. Compare these cluster treatments to those of Michael Porter (Porter 1990; November–December 1998).

The ISRN has proposed research projects and/or conducted projects on the following economic content sectors and spatial clusters across Canada:

- biomedical: Ottawa, Toronto, Montreal, Vancouver, Calgary;
- multimedia: Toronto, Montreal, Vancouver;
- culture industries: Toronto, Montreal, Vancouver;
- photonics and wireless: Ottawa, Waterloo, Calgary, Quebec;
- telecom equipment and services: Ottawa, Atlantic Canada (Halifax, Saint John, Moncton);
- wood products: BC (Kelowna), Quebec (Portneuf), Atlantic Canada;
- food and beverage: Toronto, BC (Okanagan), Quebec (Chaudiere-Appalaches), Atlantic Canada;
- automotive and steel: Southern Ontario;
- metal products: Quebec (Beauce, Mauricie).

The ISRN consists of five regional subnetworks: (1) Ontario Network on the Regional Innovation System (University of Toronto); (2) Program of Research on Management of Innovation Systems (University of Ottawa); (3) Quebec Network on Regional Innovation Systems (Université Laval); (4) Sub-Network on ICT-Intensive Innovation Systems in Atlantic Canada (University of New Brunswick); and (5) The InnoCom Sub-Network of Innovation Systems Research Network (Simon Fraser University).

The researchers of the ISRN have published the results of their research projects in a number of papers and readily available books (Holbrook and Wolfe 2000; 2002; Wolfe 2003; Wolfe and Lucas 2004; also compare these with other books that have elaborated on regional information systems – i.e. refer to Cooke *et al.* 2004; Morgan and Nauwelaers 2003).

This informative body of empirical research findings has enabled theory development, and with high value for practicing regional and urban planners. The lessons being derived from this research seek to:

> Identify best practices that are intended to inform policy formulation;
>
> Develop guidelines to enhance learning and governance; and
>
> Clarify the appropriate roles that can be played by local institutions such as universities and public research laboratories.
>
> (Wolfe and Gertler n.d.: 5)

Kitchener-Waterloo Region

The work of innovation systems researchers offers practicing regional planners a valuable body of cases from which a wide range of operational strategic and planning lessons may be derived, tailored, and inspirational to understanding the new and continuously changing circumstances of the global knowledge economy and network society. David A. Wolfe's five Ls referenced above encompassed many of the initial findings from the innovative systems research into various case studies from across Canada. Many of the findings corroborate some of the prevailing economic development theories – e.g. clusters. However, the empirical research of some of Canada's city-regions also found otherwise (Maskell and Kebir 2005). For example, in the case of the Kitchener–Waterloo, Ontario, area, it was learned that little business interaction occurs between such firms as producers, customers, and suppliers. Rather, the region's firms "compete locally on a global level," rather than interact and compete among the local producers, customers and suppliers, as is the conventional expectation from cluster modeling (cf. to the flow city concept, Graham 2002). In the Waterloo area, innovation was found to be "driven primarily by global customers in conjunction with in-house R&D departments, rather than by local competition" (Bramwell *et al.* May 12–15, 2004: 35). The linkages among the region's fastest growing firms have been represented on a large poster, referred to as a "techmap" (see Pricewaterhousecoopers 2001).

These Waterloo findings, in part, were due to the region's particular history. Refer to Part V below, section on Relational Planning Concepts, A–Z; see Path Dependence, pp. 211–12. The University of Waterloo's world-class computer-science capacity ultimately resulted from the social capital environment that was developed from the nineteenth-century industrial climate of the Kitchener–Waterloo area. Refer to Part V, section on Relational Planning Concepts, A–Z; see Social Capital, p. 218. Similar to present times, the Kitchener–Waterloo area's early enterprising industrialists needed a reliable supply of talent. They saw to it that educational institutions were developed to meet this need. They also had ensured that the then contemporary infrastructure – e.g. rail connection to Toronto and low-cost hydroelectric power – complemented the skilled personnel. This combination of education, enterprise culture and innovation ultimately evolved into the University of Waterloo's innovative cooperative education program, and in time, as the regional economy evolved into one based on science and technology, the computer science assets of the university emerged and grew. In the early stages of the evolution of this technology-stimulated region, the university followed from the industrial requirements and clustering, rather than the conventional-wisdom expectations that have been generalized from the early stages of Silicon Valley's development path, where there were strong forces of university-led development, with industries spilling over from university-based research. Portland, Oregon, discussed on pp. 82–3, also is a case of a high-technology cluster leading development, with the local university community following (Mayer February 17–20, 2002a; July 8–13, 2003). Analogous to the Waterloo region, the Portland region also has had its evolutionary and interconnected industrial relationships mapped; compare the two "maps" Mayer (May 2002b) and Price-waterhousecoopers (2001). The maps of these two cases represent the particular local regional relationalities of industry-led high-technology cluster development, in contrast to university-led clustering.

Today, the Kitchener–Waterloo region is known, branded and marketed as Canada's "Technology Triangle." The Triangle hosts RIM Research in Motion, the maker of the hand-held wireless device, the Blackberry (DeRuyter 2004). The website of Canada's Technology Triangle should be reviewed. It is an excellent example of many of the relationalities and linkages that are needed to be successful in forming and promoting an activist high-technology regional cluster (see Canada's Technology Triangle, n.d.). The website portrays the region's corporate sponsors; it identifies the participating regional and local governments; it lists the area's development awards for being competitive and fast growing; it provides research reports and local statistics, and so on. The Website, in brief, offers a one-stop electronic shop of the initial information needed to make location and development decisions.

Maritime Region

Atlantic Canada's Maritime region represents another valuable source of stimulation for relational planning innovation and planning lessons. The region has the history and characteristics that parallel many other remote areas around the world – cf. Upper Peninsula of Michigan, US. It had known economic success, yet external forces pushed it to the margins. Refer to Part V, section on Gottmann Concepts, A–Z; see External Change. Confederation in 1867 and the shift of Canada's economic and population centers of gravity westward were some of the changes external to the Maritime Provinces that moved them to the periphery of national economic development. One of the region's provinces, New Brunswick, a "lagging region," has been researched using the innovation systems approach (see Davis and Schaefer 2003). They assessed the development of the ICT sector of the province. They concluded that the young, small and fast-growing New Brunswick ICT sector is in the process of transforming the lagging region into a "dynamic region." Further, they identified five development pathways in the ICT sector:

> (1) Individual entrepreneurial start-ups, (2) corporate entrepreneurial start-ups, (3) transplants, (4) foreign servicers of local accounts that are attracted to the local market, and (5) development of exportable ICT products and services as a complement to a core non-ICT business. Each pathway implies different kinds of relationships with customers, suppliers, and supporting institutions, and a different orientation toward local, regional, or international markets. [Numbers added]
>
> (Davis and Schaefer 2003: 155)

These and other findings capture the dynamics of the early-stage development of the ICT sector. The authors have observed that next-stage growth of the sector requires "the development of deeper and broader export capability" (Davis and Schaefer 2003: 156). They also called for further research into the differences among the firms and into the factors in the local milieu that support firm formation.

Difference and Similarity

We have been trained in scientific method to look for similarities. However, both the Waterloo and New Brunswick examples illustrate that distinctive factors are associated with regional and local development. Further, development does not always conform to the prevailing development models and theories. Since all development ultimately is local, difference among the development of city-regions is the important issue to be remembered and the important lesson to be learned. It underscores that the forces of globalization are believed to have a homogenizing effect. The planning for regions and localities in the global knowledge economy and network society, to be effective, must be tailored to the special and particular qualities and assets of the place. Refer to Part V, section on Relational Planning Concepts, A–Z, see Uniqueness of Place, p. 219.

Similar to other developed national political economies around the world, Canadian regional planners too are struggling with regional planning that is more appropriate to the realities of the new city-regions. In order to meet these new conditions, Hodge and Robinson (2001) have proposed a new paradigm for regional planning. These changes for the future of regional planning in Canada call for shifts in content and focus; methods and procedures; administration and governance; and participation.

United States

In contrast to much of Eastern Asia – for example, in North America – the United States political economy especially relies much less on government initiative in extending digital infrastructure. For example, the culture and policies of the US and its states rely instead on private-sector service-provider corporations and economic market forces to provide broadband services and Internet access. In this relatively unregulated political economic environment, each of the 50 states and the corporate for-profit ICT service providers to those states represent an inherent diversity of markets, different levels of supply and demand, ability to pay and great variability in the spatial distribution of digital infrastructure, its access and usage.

The last four or five decades have seen a marked weakening of regional planning in the US. Rather than "planning," what actually takes place under this label nowadays are largely review and comment functions with little in the way of regional multijurisdictional collaboration around shared visions being planned and then realized through thoughtful rational intentionality and plan-implementation. Within the US, this heavy reliance on a market approach to digital development produces patterns of significant disparity and unevenness in the distribution of and access to digital infrastructure such as high-speed broadband and Internet connectivity. A result is measurable variation in Internet usage (Wilson *et al.* 2004). There is no standard national public policy commitment from place to place for government to ensure universality for US citizens to access the Internet and therefore for users to have a common expectation for using digital infrastructure. A potential benefit of this kind of public policy environment at subnational, local and regional scales, therefore, is that there is a great deal of opportunity for individual communities in the US to develop comparative advantages within national development, and even within international, competitive development environments. Within the context of the Canadian political economy, generally there is more reliance on government leadership in digital development, but somewhat comparable to the US, there also is a great deal of political-economy autonomy sub-nationally among the provinces, the sub-provincial regions and their respective approaches to digital access, usage and development.

Your Region is Not Silicon Valley

Historically, Silicon Valley, California, has been the often-cited model for high technology-based development (Saxenian 1994). For years, many communities around the world have sought to emulate – indeed, frequently to duplicate – "Silicon Valley development" (Rosenberg 2002). While there are many lessons to be learned from Silicon Valley's development, trying to clone Silicon Valley, or Singapore, or Stockholm, or Waterloo, or any other place, is much less effective than for each community to research pertinent best practices elsewhere, and to prepare and then to proceed to invent its own development plan and approach to planning (cf. O'Mara 2005). In order to be suggestive of a range of planning approaches that might stimulate self-learning by planners, Part IV of this book includes a number of planning scenarios at different scales that are written to illustrate various regional and local planning strategies. Rather than discuss the knowledge-economy context of the US here in one part of the book, the state of Michigan and its city-regions are used throughout the book to demonstrate many of the relational planning issues and empirical settings that frequently are faced by practicing regional and local planners in the US part of the global knowledge economy and network society. Part of the recommended ongoing process of anticipating development issues of the future for one's own region and locality, Silicon Valley specifically, and California in general often are bell-wethers for science, technology and work processes, among other possible innovations, that later might diffuse and become more pervasive (e.g. Markoff and Richtel July 3, 2005; Hanak and Baldassare 2005; Zhang and Patel 2005).

The Metropolitan Region of Portland, Oregon

From a professional regional planning perspective, when scanning the contemporary US planning-practice landscape, the city-region of Portland often emerges as a best practice case. Its innovative relational planning approaches are noteworthy, especially within the context of regional planning in the US today. Portland's particular planning approach and its planning culture have been well documented, and thereby represent a useful object lesson for those professional planners and citizen planners who want to be aware and to stay open to learning new planning behavior. Heike Mayer mapped the genealogy and evolution of the Portland high-technology industry cluster; it represented, on a large poster, the origins and relationships of the companies that formed, grew, and merged and operated to produce an important high-technology cluster, and all without a major research university. In this case, it was two important companies, Intel and Tektronix, that provided the cluster formation with the talent and research and development that conventionally is attributed to the local research university (Mayer May 2002a; February 17–20, 2002b; compare Institute of Portland Metropolitan Studies n.d. to Pricewaterhousecoopers 2001). For insight into Portland's planning, Ozawa's edited book, *The Portland Edge: Challenges and Successes in Growing Communities* (2004) is a "must" read. It provides informative insight into the context for Portland's different approaches to such planning issues as equity; regional planning and development control; a local and innovative knowledge economy, but without a major research university; the resulting housing market; livability and quality of life; progressive civic behavior and community participation; downtown development; transportation planning for the metropolitan area; environmental sustainability, and many other development issues with relationality to the city-region's special qualities. However, even in the context of a planning best practice and benchmark case, this city-region does not necessarily out-compete its peers. Recently, Portland lost an important biotechnology development project to Seattle. The story of a hometown Portland development company decided to build this project

> in Seattle, not in Portland, reveals much about the way the two cities conduct business. In Portland, powerful city planners set the agenda for development. In Seattle, which has the advantage of an established foothold in the biotech industry, developers are given more control – provided their goals match with the city's.
>
> (Rivera October 24, 2004: B1)

One of the many lessons that may be derived from this case is that planners and planning in the US may need to be more agile and flexible than their counterparts in other countries and political economies, if their planned strategies are to be realized.

Michigan and its City-regions as a Surrogate for the United States

From the perspective of regions within states, and in the case of the state of Michigan, for example, it is clear to us that the people and their planners need to pay much more attention to, and concern for international economic competition and the approaches that are used by global competitors. It is imperative for US grass-roots leaders, stakeholders and their planners to earn an understanding of the success factors that competitors and their planners use (Williams and Stimson 2001). As noted earlier in Part II, the role of benchmarking and the study of best practices is essential in gauging the competitive environments external to the locality and its planning region.

In the US, there has been a long tradition of the public and the electorate, over time, swinging from belief in market-led development in contrast to government-led interventions for development and vice versa. Increasingly, it is clear that these approaches to development are not in opposition. Rather, the emphasis one way or another shifts over time and from place to place within the regions of the US. In today's global knowledge economy and network society, what is likely to be most effective for US ICT-enabled regional development planning is complementarity, coordination and alignment of purpose among a locality's governments, the market, the business sector, the nonprofit sector and individuals. Each of these sectors and actors has an important contribution to make to regional and local development, especially over the phases or life-cycle of a development planning process. This mix of roles by the respective key development actors is likely to be effective because this approach, when executed competently, is representative of the wide and diverse base of a locality's principal development stakeholders and their respective interests.

Eastern Asia

In the Huggins *et al.* (2004) *World Knowledge Competitive Index 2004* report, Eastern Asia ranked third among the three major technology-economic regions of the globe, behind North America and Western Europe. Of the 125 regions worldwide that were ranked, only 25 were in Asia; 17 of these were in Eastern Asia, and 8 regions were in other areas of Asia (Huggins *et al.* 2004: 6). As noted earlier, the greater Tokyo region had the Tokyo and Shiga regions as the only Asian regions ranked in the world's first 50 knowledge-competitive regions. Japan has 7 other regions ranked among the next 50 regions globally. With their global ranking listed in parentheses, other Eastern Asia-Pacific regions included: Singapore (74); Australia with 3 regions – Victoria (79), New South Wales (83) and Western Australia (85); Ulsan (101) and Seoul (109) in South Korea; Taiwan (102); Hong Kong (106); New Zealand (108); and China, with three additional regions rounding out the remaining Eastern Asia-Pacific ranked world knowledge competitive regions.

Japan, Singapore and Asia

In modern historical context – for example in the 1980s – Japan was one of the earliest leaders of the world in seeking to harness the development potential of "high technology," especially by means of its technopolis strategy – for example, see Tatsuno 1986; cf. Morris-Suzuki 1994. Another early information technology (IT) and service-economy pioneer of the era and of the Eastern Asia region was the city-state of Singapore. Its high priority for planning and making investment in IT in the 1985–6 period was in part a function of responding to a contemporary worldwide economic recession. As a city-region and country, Singapore has been particularly noteworthy as a benchmark case of one of the world's earliest and continuing practitioners of innovative strategic planning for digital development and, more recently, for intelligent development. Further, Singapore's early planning of digital development innovations operated to stimulate analogous development innovation by Singapore's competitors and near-neighbors in Southeast Asia (Corey 1998; 2000; and cf. Ho *et al.* 2003).

Singapore's Digital Development

Government leaders of Singapore, even in the 1960s, foresaw that they should consider technology and science as drivers and enablers of the country's economic development. Because of Singapore's long-term reliance on, and commitment to strategic development planning, and because of its early IT policy innovations, we have been able to research, monitor and assess the country's urban technology progress for nearly a generation (Corey 1987; 1991; 1998; 2000; 2004; Corey and Wilson 2005b; 2006). Singapore's planned change actions and their implementation in ICTs and science applications invariably have led such economic performance for Southeast Asia, and have been pioneering at both the Asian and global levels. Singapore leaders have worked continuously on such planning as part of the country's planned policies to maintain a competitive edge in the global knowledge economy and to its more immediate regional economies. Because of the city-state's small population and tiny national territory with few natural resources, its government leaders and its people saw little choice but to be innovative, creative, and to plan to be several steps ahead of the competition. While the culture, its dominant value system, and the truly unique political economy of the place combine to defy emulation and re-potting of its intelligent development planning style and approach to locations elsewhere, Singapore does model for us many of the behaviors and practices that are likely to result in knowledge-economy development "success." Singapore's technology and knowledge-induced planning has been integrated effectively into its other planning functions, from physical and transportation development, to economic development, to education, to housing, and among other functions, into its planning objectives for community identity and culture-development planning. Singapore's plan-making usually is followed up with well-coordinated, clearly prioritized, and well-funded programs of plan implementation. It has had to work hard on improving the entrepreneurial and risk-taking local culture by instituting many initiatives to promote creativity and innovation.

The Singapore government continues to address these challenges (Rodan 2004). The next iteration of Singapore's technology strategy was released recently as the ten-year Infocomm Technology Roadmap (Infocomm Development Authority, March 2005: 4–5). The Roadmap is the city-state's new ICT master plan. It is labeled "Intelligent Nation 2015." This is Singapore's first ten-year long ICT plan. It seeks to continue to keep the country competitive in future by means of accommodating three waves of change: (1) nanotechnology and biosciences supported by computing; (2) "advanced ubiquitous broadband technologies and short range wireless communication technologies"; and (3) smart sensor technologies (Infocomm Development Authority March 2005: 4).

Singapore's Intelligent Development: Culture and Creativity

Singapore is noteworthy not only for its continuous planning and for the implementation of digital development, but also because it has made major progress in intelligent development. The city-state long had been criticized for the extensiveness of its urban renewal and the uniformity of its large high-rise housing estates. Visually and in terms of the quality of life, the city was said to be sterile. However, in just a decade, Singapore has made noteworthy advances in creating and putting on an upward trajectory – a vibrant arts, media and culture scene – all within the context of developing a more attractive and beautiful physical and cultural environment (Corey December 18, 2003). This activation has taken hold across the cultural spectrum, including literature and poetry; the arts; performance; popular music; design; and, to some extent, architecture and urban design (Urban Redevelopment Authority n.d.), among other creative endeavors. Given the ethnic diversity of Singapore's population much of the growth of the cultural environment is characterized by a fusion of the cultural richness of the Southeast Asian region (Savage 2000).

It is instructive to examine how Singapore has gone about activating its planning vision to transform itself into a center of excellence in culture and the arts (Ministry of Information, Communications and the Arts August 2002). This vision was constructed as part of a range of activities that is intended to be directly supportive of enhancing the country's creativity and competitiveness within the context of the global knowledge economy and network society. Having a vibrant cultural and arts scene is conceived as a factor in attracting, retaining and growing creative and satisfied knowledge workers and citizens. Five conditions have been identified to realize this vision: (1) sustained economic growth; (2) patronage of the arts; (3) developing the arts software; (4) government funding; and (5) increased appreciation of the arts (Savage 2000). Further, the establishment of such culture-support conditions can serve to generate a variety of support functions and "industries," ranging from arts education to arts management, to arts packaging to the engagement of the corporate, the voluntary, and philanthropic sectors so that they may play their critical parts in advancing life-culture and the arts. Investment priority in science and technology human capital, and priority for culture and arts development combine to form an intentional strategy to enhance Singapore's environment of innovation and entrepreneurship.

As Singapore's population ages, as leaders change, as forces for greater political freedoms arise, as new generation leaders emerge, and as other external and internal dynamics generate changes, new conditions and perceptions by Singaporeans, development lessons may continue to be derived from the Singapore experience. It is one of the globe's few national economies that has planned to embrace fully the knowledge economy potential for its future development.

Eastern Asia: Evolution and Maturation

Long-term analyses and reflections on Singapore's experimental approaches to ICT- and science-induced planning have enabled us to observe how near-neighbors have been influenced (Corey 2000). Our observations also have permitted us to draw conclusions and come to forecasts about likely technology-based development in other parts of Southeast Asia beyond Singapore. For example, a number of other countries in the region have sought to adopt similar planned strategies. While some technological leapfrogging over development stages has been demonstrated, external and internal dynamics have functioned collectively to maintain stable disparities and rankings in the countries' digital development levels and intelligent development levels (Corey 2004; Corey and Wilson 2005b; 2006).

The International Telecommunication Union (ITU) conducted a 2002 study that generated digital access rankings of countries around the world (International Telecommunication Union November 19 2003). The results revealed digital divides and access differences among countries. The measures used by the ITU to construct the concept of digital divide represented it as a function of affordability and education, as well as limited digital infrastructure. The ITU's Michael Minges, who was responsible for the 2002 study, "Digital Access Index," observed:

> The four Asian Tigers (of Hong Kong, Korea, Taiwan and Singapore) have made the greatest progress over the last four years. English-speaking countries are falling behind. He contrasted government promotion of Internet access in Asia with the tendency of Anglophone nations to leave the job to the private sector.
>
> (International Telecommunication Union November 19, 2003: 1; Williams November 20, 2003)

Among these four Eastern Asian economies, there is a great deal of variation in political economy approaches and resultant policies. For example, there is Hong Kong at one end of a political economy and policy spectrum, and Singapore at the other end, with the Republic of Korea and Taiwan in between. Hong Kong represents a highly free-market political economy; Singapore illustrates a political economy with a high degree of government activity in the market, while also promoting and constantly attempting to stimulate private entrepreneurship, innovation and creativity both at home and for application abroad.

Latecomer Policies: The Case of Taiwan

Taiwan's policies for responding to globalization forces is informative to planners and their stakeholders who face the situation of having to compete economically without the benefit of having technologically cutting-edge firms either in manufacturing or services. Since such regions already are engaged in latecomer product and service development, what are the future development options for the region and locality when even these activities and jobs are moving elsewhere or are being eliminated because of the substitution of technology for labor? Amsden and Chu have developed a model and policy analysis for consideration and use in locations with firms that are struggling with just these latecomer and mature product life-cycle issues. Their policies analysis demonstrates the pertinent assumptions to make and to avoid policy approaches that are less valid for the particular industrial stage of the region and its specific mix of industry sectors, both existing and planned (Amsden and Chu 2003: 161–76). Their policies analysis is a model for how regional and local planners must parse out the appropriate development product and industry stages and apply the mix of government and private interventions and investments that are most applicable to the local region's capacities and competitive advantages. It is an excellent case of translating theory – e.g. product life-cycle and industry life-cycle – into strategies for sensitive public and private policies that are based on alliances among government and the area's firms. Refer to Part V, section on Relational Planning Concepts, A–Z; see Industry Life-cycle Model, pp. 203–4 and Product Life-cycle Model, pp. 213–14.

Asian Relational Theory and Philosophy

The highly dynamic and diverse technology and economic environment of the region, in part may be attributable to Eastern Asia's ancient cultural and spiritual traditions. These traditions have helped to spawn distinctive theoretical and philosophical innovation (Lim 2003). Somewhat analogous to recent emerging scholarly interest in relational approaches in Europe (Graham and Healey 1999), there has been interest in Asia in practicing new economic geographic relational thinking, especially relative to the Asian development context (Yeung March 13, 2002; 2003; and 2005). Henry Wai-chung Yeung has called for us to be critical of economic geography theories on Asia from sources outside Asia, including such applications from the Anglo-American contexts (cf. Kunzmann July 3, 2004). Yeung and Lin have observed that new development approaches used in Asia should be theorized and added to the global discussion and debate about the economic, spatial and development dynamics of the knowledge economy and network society (Yeung and Lin 2003; and cf. Yeung 1998).

Indeed, planning and public policy discussion in Asia has a related distinguishing characteristic that should be noted. An explicit spiritual dimension may be discerned in public policy and planning debate in several Asian nations – e.g. Korea, Sri Lanka, Indonesia, Thailand and Malaysia, among others. Planner Gill-Chin Lim exemplified these characteristics in his advocacy of "life-culture": "Culture in which life of every human being is cherished, culture in which life sustaining human activities are carried out with a sense of living in harmony with others and nature" (Lim December 19, 2003: 3).

In Singapore, related debates long have taken place under the rubric of "Asian values." The teachings of Confucius are widely referenced both within and outside of Asia. However, as another indicator of the diversity of social thought of Asia, Gill-Chin Lim has called attention to the fact that five major schools of thought from Asia influence governance behavior. He was concerned especially with the role of trust in good governance. The various schools of Asian social thought are: (1) the Confucian School; (2) the Taoist School; (3) the Legalist School; (4) the Warfare School; and (5) Buddhism. Influenced by the tenets of these Asian philosophical schools, Lim recommended the following actions for consideration in designing and implementing a strategy for good governance:

> Invest more in education.
>
> Reform the public sector, with special attention for instilling ethical competence.
>
> Reform the business sector, to reeducate its human resources, restructure its governance system, and manage under high standards of ethics that is codified, reviewed and enforced.
>
> In order to save money and protect human life, all sectors must maintain technical codes at high levels.
>
> Civic organizations should develop the capacity to monitor, educate and disseminate information on governance behavior and results in and to the public and business, as well as maintaining their own ethical standards.
>
> The mass media need to attend to the importance of culture and trust relationships as critical community-development ingredients in public and private-sector governance.
>
> Since individual persons comprise a nation's governance culture, national and local campaigns should be conducted across all sectors of society that are effective in growing trustworthy individuals and a society.
>
> In order to achieve improved levels of trust in governance beyond the nation, an international agency, such as the United Nations, or multiple agencies should take up the role of governance reform for the purposes of strengthening government and business globally.
>
> (Lim December 19, 2003: 29–30)

The criticality of values, culture and history in development too often are neglected in regional and local planning and plan-making. In today's climate of widespread mistrust of relatively unresponsive and indifferent governments, frequent instances of corporate corruption, as well as cheating and dishonesty at the level of the individual, good governance is unlikely to be successful without close attention to these recommendations from Gill-Chin Lim. For a discussion of the role of tradition in development, refer to Goulet 1983: 19–20.

Asia or Many Asias

One of the many qualities of Asia, in relative contrast to the national economies of North America and Western Europe, is the diversity of national development policies and approaches. This makes for a rich source of learning. This large technology-economic region includes several of the earliest digital development policies planning cases – e.g. Japan and Singapore – and, more recently, the world's largest – i.e. the People's Republic of China. With this relatively long period, and with this diverse range of national policy frameworks, this huge region should be monitored continuously in order systematically to learn from the development and planning innovations, best practices and benchmarks that are constantly being invented there. Possibly because of the national-level diversity of development, this region long has planned and implemented mega-projects and large infrastructure projects. Such efforts are seen to help establish and maintain national and local identity. This proclivity to think and act big has produced a receptive environment, often with the requisite resources and priorities behind such projects to plan and implement tallest buildings, largest bridges, longest tunnels, most modern airport complexes, fastest trains and so on. The interweaving of these various tendencies – i.e. different national identities and their respective primate city-region hubs – has given rise to political and economic interests in seeking to better connect, network and complement these differences. Such linkages are seen as good for business and for civil relations among the national cultures and economies. As a consequence, "growth triangles" (Thant and Tang 1996) and "technology corridors" (Corey 2000; 2004a) and other planned transnational corridor efforts (Choe 1996; 1998) have been planned and actually built, with even more mammoth visions now on the drawing boards (Park April 25, 2003). These kinds of dramatic and highly integrated electronic and physical development projects (Rimmer 1994; Morris-Suzuki and Rimmer 2000; Lo and Marcotullio 2000) stir imaginations, attract investment and stimulate global-level excitement. Both the positive and the often less obtrusive darker sides (Bunnell 2004) of such development need to be followed, with lessons drawn and appropriate solutions achieved for one's own regional and local planning. Too often these actors in the West have a tendency to view Asia more monolithically and uniformly than is the case; this inclination should be avoided. The many prisms of Eastern Asia's development policies and planning cases have much to teach regional and local planners, their stakeholders, and citizen planners across the globe. The great reverence that many Asian cultures have for education and respect for family are values that merit close attention within the regional and local development context in today's global knowledge economy. Further, the highly competitive and productive economies of Asia, coupled with many familiar and tight connectivities from Asia to other parts of the world and return, demonstrate relationalities that are not duplicated elsewhere. In short, these qualities comport nicely with the two key contemporary drivers of regional economic growth that were identified in Part II above by Plummer and Taylor – i.e. (1) human resources or human capital; and (2) an enterprise culture (Plummer and Taylor 2001a; 2001b; 2003). The many prisms of Eastern Asia's development policies and planning cases have much to teach regional and local planners, their stakeholders, and citizen planners across the globe. The need for the attainment of these qualities is reiterated throughout this book. The leaders and planners of each place must draw on their own particular traditions, cultures and local assets in order to be successful in this task.

Western Europe

Western Europe has provided technology leadership, especially in the policy sphere, for many years. France developed Minitel, one of the first large-scale ICT developments that preceded the Internet by a decade. The European Union adopted a broad perspective with the commissioning of the Bangemann Report on *Europe and the Global Information Society* in 1993 (European Commission May 1994). The report took a continent-wide perspective that allowed policy to be developed to capture the benefits of ICT developments at a time when many were unaware of the changes to come, and were unwilling to act on the emerging technologies. The early policy focus on implementing ICT in Europe allowed many communities to develop e-commerce, e-government, and science and technology initiatives.

With the expansion of the European Union and attention on EU policies, there is naturally an expectation that large-scale global regions offer advantages. At the same time, the focal point for a lot of economic development remains at the local level. Large-scale political and economic re-alignments such as the European Union and the North American Free Trade Agreement (NAFTA) have provided changed policy frameworks for the rise of the city and region. Sassen (1991 and 1994) showed the significance of global cities as engines of growth, while Brenner (2004) and McNeill (2004) point to the role of smaller-scale European cities as entities for change. Strategic planning attention has been directed at the large-scale economic development policy framework of the European Union, and local planning has stimulated the advanced urban regions of Europe to drive economic progress.

The *World Knowledge Competitive Index 2004* has ranked more than 40 Western European regions of the total 125 worldwide regions analyzed in the study by Huggins *et al.* (2004). Their global rankings are listed here in parentheses. The Stockholm, Sweden, region (15) and the Helsinki, Finland, Uusimaa region (19) are the highest-ranked regions in Europe. While having lower ranked regions, Germany has Europe's largest number of regions at 10; Italy has 6; Sweden has 5; the United Kingdom has 4; and the Netherlands has 3; the historically and culturally related and highly interdependent Nordic countries (of Denmark, Finland, Norway and Sweden) have a total of 10 ranked regions. The level of these rankings of competitive knowledge-economy regions and the numbers of such regions, in general terms, are indicative of the relatively even spatial distributional patterns of knowledge-economy city-regions across Western Europe. The cumulative rankings, in general, place Europe's regions below North America's US regions and ahead of Eastern Asia's regions.

Europe offers many examples of innovative cities and regions that have leveraged existing resources and combined them with innovation and thoughtful public policy. Often, cases focus on major areas such as London and Southeast England, Randstad, Bavaria and the Ile de France as examples of regional development in a knowledge economy. For local planners, however, there is greater value in examining how smaller cities and regions have prepared for economic change in a global economy.

Skövde

Skövde is a city of 50,000 people in central Sweden, with an economic tradition in manufacturing and reliance on two Volvo engine plants for employment. With the changing economy, the city is looking to its university, founded in 1977, as a source of potential growth utilizing expertise in ICT and life sciences. The Gothia Science Park was established in 1998 and now has 40 companies and 150 staff (Lorentzon June 2005). The science park exemplifies the ability of small communities to utilize knowledge resources for new economic functions. It is important to note, however, that the existence of a university alone is not sufficient, as it takes vision and leadership, as well as strong social capital among leaders to develop such a cluster.

Sophia Antipolis

Sophia Antipolis is a research cluster located near Antibes in southern France. The name connects knowledge with the Roman name for Antibes – Antipolis. Started in the 1970s to take advantage of empty space in a location able to attract knowledge workers, Sophia Antipolis was a successful technology cluster. It was built on links to universities, firms with global networks, research centers and a high-quality amenity value (Van Looy *et al.* 2003). The decline of the computer industry and the shift to innovation as a primary function hurt the area as it grew on external links rather than the synergy of internal links that are essential to the innovation process (Lazaric *et al.* 2004). The early success of the cluster was built on the presence of essential inputs to knowledge creation combined with an excellent location. As the knowledge generation process changed, however, the cluster lost some of its ability due to a lack of intra cluster linkages and capital on which to base new research and development.

Benevento

Benevento is a province in the Campania region of Italy, a region traditionally challenged to find employment for its residents. The establishment of the University of Sannio in 1989 became a catalyst for recent development. Among new initiatives that link the university, province and business are a center of excellence for software development and the Mediterranean Agency for Remote Sensing that both fit small but important niches in applied science and technology (Mucerino and Paradiso June 2005).

Lessons from Europe in Strategic Spatial Governance

Stimulated by the European Spatial Development Perspective (Faludi and Waterhout 2002), planning researchers of European city-regions have derived five lessons from the European experience to date in experimenting with ways of "strengthening regional identity and developing new forms of regional collaboration." To paraphrase Albrechts *et al.* (Spring 2003), the planning lessons that they have noted are: (1) strategic spatial planning initiatives take many different forms; they need to perform different governance functions depending on the particular local context; (2) such initiatives both release forces of innovation and can reinforce the status quo; (3) developing spatial strategic plans within a governance context requires technical analysis – i.e. the "development of spatial logic and metaphors" – and, in the end, organizing images need to be invented and adopted locally – e.g. the Dublin Belfast Corridor in Ireland; (4) "appropriate institutional arenas" must be created for regional spatial development initiatives; they note that often it is necessary to create "a new kind of governance culture"; and (5) strategic spatial planning "initiatives benefit from the existence and acceptance of a strong role for the state and a strong political consciousness of place identity" (Albrechts *et al.* Spring 2003: 127–8). When preparing to engage the global knowledge economy and network society from the perspective of one's own regions and locality, strategic planning that takes into account these key relational spatial planning lessons from the European experience is likely to prove beneficial. Also, refer to Healey's forthcoming book on relational planning.

The influence of the ancient world on our collective understanding and studies of urban dynamics and relationships was discussed in Part II (see pp. 38–9). From the policy perspective of the more recent modern world, of national information technology strategic planning, France was the earliest country to engage the challenges of the digital development era. For example, in 1978, the President of France, Valery Giscard d'Estaing, was sent the landmark report, *The Computerization of Society* (Nora and Minc 1978). Europeans have been the world's most significant recent contributors to modern and contemporary regional planning for the global knowledge economy and network society. This is a function of several factors. Two of the more instrumental factors include a long-standing valuing by Western Europeans of the welfare-state political economy, and the influence of an overarching set of policies and programs of the European Union (EU) that provides a digital development framework for a diversity of EU countries and their respective subnational regions. The European Spatial Development Perspective (ESDP) in particular affects the European commitment to and mindset toward regional and local planning. This does not necessarily mean, however, that there is a homogeneous or uniform approach across the continent to regional development planning. Because of these factors, it should not be surprising that noteworthy theoretical and evaluative research and writing have been conducted by Europeans and empirically in Europe, as has been referenced throughout the book and especially in the case of Patsy Healey and her relational planning researcher colleagues in Europe. In the end, one can trace an evolution of key concepts and relationships both for the study of cities and regions, and for the planning of cities and regions – i.e. a transcendence of intellectual leadership from (1) the ancient classical world of the Mediterranean, to (2) the modern and contemporary world of Europe and selected cases of evaluating individual city-region relational planning examples, to (3) studying the networking issues associated with sustainable management of European polycentric (cf. Healey May 2001) large city-regions that exhibit concentrated deconcentration – for example, Sir Peter Hall and a network of urban scholars across North and West Europe are implementing a multi-year research project examining global economic and knowledge flows from major cities to smaller cities within their spheres of influence for seven areas of North-West Europe (Hall *et al.* July 4, 2003). These city-regions include: Southeast England; selected city-systems of the Netherlands; the Rhine-Main region of Germany; the Dublin–Belfast corridor region of Ireland (Urban Institute Ireland n.d.); the Paris region; the Metropolitan Region of Zurich; and the Rhine–Ruhr region. This scholarly collaboration is a descendant of Jean Gottmann's Alexandrine model of the ancient classical period, and, among other relations, will reveal contemporary local political-economies that can function as grist for the mill of urban and regional planning strategizing for the future (Hall *et al.* July 4, 2003). This kind of important research investment by the EU into the city-region dynamics of the knowledge-economy can be expected to advance significantly our contemporary collective wisdom on these emerging complex issues.

Globalization Influences on Contemporary Planning Practice

Globally, from the perspective of technology-enabled planned development, the world's core technology-economic regions include North America, Eastern Asia and Western Europe. For the purposes of deriving development lessons for future strategic planning, these regions merit our ongoing attention and reflection. The political economies of these regions are instrumental in determining the policy and planning practice environment of the countries of each region.

The resulting differences and similarities within and among the core regions are rich and deep; analysis and comparison of these realities can stimulate innovation and fresh approaches, both for the future study and future planning of cities and regions.

It should be emphasized that we advocate deriving development-planning lessons from anywhere in the world. The developing world is rich and deep in charting its own diverse pathways of digital and intelligent development. China and India in particular are, and will continue to be great sources of experimentation and learning. As the regions of the Middle East, Africa and Latin America continue to engage the various forces of globalization, expand digital infrastructure and incorporate digital development into their strategies and policies, they will contribute much new actionable knowledge. We have called for the creation of a global planning-practice learning community that will work to make heretofore inaccessible sources of local planning lessons accessible. Because of language, cultural, financial and other barriers, it is difficult to learn from each other:

> Our dream is for a global free market of *innovative* planning ideas and practices. Such a market calls for openness and a recognition that all planners have something to learn from, and share with each other. One hopes that one of the themes for debate and emulation are the invention of effective planned actions that result in the conservation and preservation of the best characteristics that confer identity and distinctiveness on our respective regions and localities.
>
> (Corey and Wilson July 2005: 6)

As planners around the world struggle with the influences of globalization, one reaction might be to isolate one's self from these new complexities and uncertainties. We counsel otherwise. By being open to new ideas and being willing to be stimulated to innovate and experiment locally, practicing planners will be serving society better, and in the process, functioning to empower their clients, constituents and local stakeholders. This, after all, is one of the fundamental reasons for the existence of the profession of planning. In a most humble and modest way, the next part of the book turns to practical ways to address these global issues locally.

Part IV
"ALERT"

A Model for Regional and Local Planning Practice in the Global Knowledge Economy and Network Society

Convergence in Place

Many of the issues, concepts and practices that were noted in the earlier parts of the book may be made to come together in the region and its localities when one seeks to mobilize this knowledge and to activate the planners and stakeholders of the community on behalf of the planned future for the area. The emphasis in this part of the book is on stimulating planning and action; the knowledge introduced and discussed above now must inform a creative planned program of implementation.

To have a good chance at being successful in this mission locally, the old legacy-planning approaches need to be examined and changed when they are found to be incongruent with the new dynamics of the global knowledge economy and network society. This means deviating from, and overcoming path dependence; changing the mindsets of the area's planners and stakeholders; developing a new and effective strategic governance structure for the region; and attending to the disparities inherent across the area's economically distressed communities. Among other initiatives, the regions need to give the highest investment priority to (1) investing in, and realizing high-quality local human and intellectual capital, and to (2) the development and the realization of a creative enterprise culture.

The discussion that follows for this part of the book is normative – i.e. the emphasis is on "should." The suggestions for planning approaches are based on, and have incorporated within them, the "what," or many of the knowledge issues that have been discussed above in each of the previous parts of the book. *It must be reiterated that these normative notions are offered more as stimuli rather than guidelines or solutions that apply universally*. It remains for you, the planner, and your stakeholders to fashion your own strategies by drawing on your particular special assets and mix of actors. Again, each region and its localities have their own mix of potentialities, problems and personalities that demand special attention and tailoring. In practical terms, the following general prescriptions seek to answer the question that community stakeholders always ask: "Okay, you have our attention, we are becoming aware; now what do we do?"

The narrative refers to “the region and its localities.” This general phasing is used to suggest to the readers of the book how they might go about engaging their own area and place in such thinking and acting. In other contexts, the phasing may refer to a specific actual region – e.g. a substate region of Michigan.

From Digital Development to Intelligent Development

Relational Planning: The Conceptual and Organizational Basis for Intelligent Development

Introduced in Part II above, Graham and Healey (October 1999) provided practicing planners with initial useful conceptual guidance on how to incorporate a new global knowledge-economy geography into their visioning and planning. Based in large part on relational theories of time and space, they argued for changes in planning practice. These changes were congruent with the new complexities and subtle dynamics of the global knowledge economy and network society. Four interrelated guiding points or principles were offered by Graham and Healey for the practice of relational planning. They advised planners seeking to practice relational planning to: (1) consider relations and processes rather than the traditional planner's emphasis on objects and forms; (2) stress the multiple meanings of space and time; (3) represent places as multiple layers of relational assets and resources, which generate a distinctive power geometry of places; and (4) recognize how the relations within and between the layers of the power geometries of place are negotiated by the power of agency through communication, interpretation and new governance behaviors.

What Do We Do Next?

Once the clients, constituents and stakeholders of the regional planner become aware of the changed dynamics and new opportunities of the global knowledge economy and network society, they invariably ask, "What do we do next?". Guided by the concepts that underpin the new planning-practice behaviors suggested by Graham and Healey, the answer to this question is to develop a collective vision for the region's future. For today's global knowledge economy and network society, it is imperative to have a substantive strategic vision toward which to direct planning action. Four operational visions, in interdependent combination, can stimulate the collective creativity required to lead to a more competitive region: (1) development; (2) digital development; (3) intelligent development; and (4) the content and locational framework of the e-Business Spectrum.

With these and other emerging relational conceptualizations newly integrated into their changed mindset, today's regional planning practitioners may move their planning leadership and facilitation beyond digital development which is focused on access to ICT infrastructure to intelligent development that is informed by best practice experience, appropriate theory – i.e. relational theory, and sound empirical analyses, thereby the region and its localities might realize more effective and competitive contemporary planning practice and development content outcomes.

Development

The end-state vision for the region and its localities should be composed of the fundamental functions and factors that are required for realizing an economically productive community that is able to compete effectively in the global economy. To be more fully productive, targeted investment must be made to reduce economic, social and quality of life disparities throughout the region and its communities. Refer to Part V, section on Relational Planning Concepts, A–Z; see Development, p. 196.

Digital Development

In order that the people and the organizations of the region might be able to conduct the business and life of the area, the required contemporary infrastructure must be in place. This means having the modern electronic infrastructure needed to network, communicate and transact the region's business. This is digital development, which is the application of information and communications technologies (ICTs) infrastructure to community economic development. Such infrastructure includes high-speed broadband and wireless technologies, for example.

Intelligent Development

Intelligent development is digital development that explicitly draws on and is guided by appropriate concepts and theory – e.g. especially location theory, economic development theory and relational planning theory, in the formulation of policy and the planning of communities for successful competition in the global knowledge economy and network society. It promotes investment in a place, thereby being productive in wealth creation, human capital development, employment formation, creation of an enterprise culture and improvement in the region's quality of life and equity. Such intelligent-development planning should focus on maximizing the value-added of a place by matching and segmenting the unique e-business functions, factors and relations of the locality and its region to the appropriate stages in the life-cycle processes of production and consumption functions. Development is "intelligent," therefore, when the best practices from theory, benchmarking elsewhere and the appropriate applications of the latest technologies are utilized fully to develop a region and its communities holistically, multidimensionally and equitably. This experience-based and theory-informed approach to planning and plan implementation has been conceptualized as "intelligent development" (Corey December 18, 2003).

Stakeholders: Actors and Roles

In order to implement the ALERT Model introduced on p. 108, there must be a mechanism that provides direction and governance for sustained strategic planning activity. Below is a discussion of some of the key kinds of issues and principles that should be considered in selecting representatives of the regional economy for a strategic planning steering group whose role is to provide local governance leadership – i.e. leadership that is in direct follow-up to the "awareness" phase and in preparation for the "talk" stage of the ALERT Model, both of which are discussed below. To understand better the dynamics and relations among the various actors and their roles, it may be helpful to conceive these behaviors as a kind of "drama" – i.e. a theatrical-like performance. Refer to Part V, section on Relational Planning Concepts, A–Z, see Drama, p. 197.

Given the multiple functions, complexities and linkages inherent in such local knowledge-economy planning, a coordinated division of labor will be required. Under the overall direction of a "strategic planning steering group," representative "working subgroups" also need to be organized to assist in executing individual components and tasks of developing the regional strategy. Typically, the steering group and its subgroups would be staffed and facilitated by the city-region's professional planners, or other related professionals, such as economic developers. For rural and remote areas, where there may be few, if any, professional planners, others may need to assume the role of leading and supporting the strategic planning working subgroups in their tasks. Such other leaders might include professional economic developers, local elected officials, nonprofit organizational leaders and so on. These professional or volunteer facilitators should comprise "planning support teams," and their chief team facilitator may function in the role of planning-support team leader. In discussing the leader role for planning practitioners, Myers and Banerjee have stated:

> In the practice of planning, we would lead local efforts to solve urban problems, lead the new dialogue about growth visions and futures, lead the building of collaborative partnerships, lead the partnerships fostering

> a new regionalism, lead international efforts for managing urban growth and development planning, and lead the campaign for urban sustainability, among others.
>
> (Myers and Banerjee Spring 2005: 128)

The guiding principles for advancing the activities of the ALERT Model come from knowledge of a body of behavioral science theory that is well tested and proven. The key is that a strategic planning steering group must be representative of the community and region for which it will advise on the area's strategic planning. The principal economic, social and cultural sectors and institutions of the region must be involved. The following are the generally recommended categories of representation for an effective strategic planning steering group of stakeholders (cf. Brody Summer 2003). It is from these categories of stakeholders that members for the strategic planning steering group should be selected:

- Business, including small and large enterprises, as well as organized labor.
- Public-sector actors such as representatives from the various levels of government, such as local, multi-jurisdictional, regional, including executive and elective office holders – e.g. local legislative or parliamentary representative(s).
- Nonprofit-sector institutions, especially education representatives at the critical levels of tertiary (including from community colleges or polytechnics), secondary, and primary/pre-school, plus area social welfare organizations, arts and other cultural groups.
- Selected professions should be considered for representation on the regional and local strategic planning organization. In particular, urban and regional planners, and professional economic development practitioners should be targeted for special inclusion. These are the professionals whose combined primary practice missions are to advance the planning and the implementation of the community's economic development. It is the mindset change of these professionals that is so critical to the early and long-term civic behavior change that explicitly relates and engages the region and its localities with the global knowledge economy. In the case of local economic development practice – for example, in order to move it from the traditional foci on land, business attraction and marketing – contemporary practice also needs to incorporate some of the critical factors that are important to economic development in the era of the global knowledge economy and network society, such as sustained investment in human capital and the creation and support of a local enterprise culture of innovation and entrepreneurship. Five reasons have been found to explain why professional economic

development practitioners at the regional level have been slow to engage the global knowledge economy: (1) institutional inertia; (2) rather than rewarding and holding accountable economic development practice at the level of the regional economy, the convention and norm is to assess success at the level of the local jurisdiction; (3) the functions of regional economies need to be reinforced and supported by institutions and measurement systems also operating regionally, rather than primarily locally; (4) new regional-level business leadership needs to emerge and prevail that emphasizes the factors critical to global knowledge economy development, such as venture capital and investment in all levels of education, including tertiary education, in contrast to traditional economic development leadership that assigns highest priorities to land development and property tax enhancement; and (5) in this era of globalization, local and regional business leadership needs to recognize that corporate leadership increasingly is being centralized nationally and internationally, thereby putting higher expectation on locally-owned and controlled business leadership to be more assertive and proactive, thereby taking up the slack and providing stronger representation of the local business community among the mix of actors involved in transforming regional governance and strategic planning (Mayer *et al.* October 21–24, 2004).

- Individuals who are opinion leaders and/or persons who are influential (and/or wealthy) and whose civic views and service to the community have attracted respect – e.g. developers, investors, media owners, etc.

The Planning Support Team is the group of regional and local professional planners or others working with stakeholders and citizen planners to provide advice and consultation on the implementation of the ALERT Model. The primary role of the planning support team is to bring external knowledge into the regional and local strategic planning process, including the scenario planning activities of the TALK phase of the ALERT Model.

The formulation and operating principles for the Regional Strategic Planning Steering Group of stakeholders should be:

- Ideally, the planning steering group and subgroups each should be 9–15 persons in size and well staffed, including by volunteer staff; odd numbered groups are best; the key issue here is to have a group that can function effectively and not be unwieldy.
- Diverse and multi-generational in nature; a blend of both younger and mature representatives is critical to the success of the group's planning process. The group needs the fresh and innovation-oriented perspectives of the younger participants, and the practical and feasibility-oriented perspectives of the mature and experienced participants. The principal ethnic and racial communities of the area must be fully represented and involved, including, where applicable, aboriginal or first-nation interests.

- Resource controllers must be at the table (or their trusted delegates); assuming a voluntary effort of cross-jurisdictional, intergovernmental, interinstitutional and multicorporate actors; there must be full involvement of representatives who can, and are likely, to bring budgetary and/or grant funding and other resources (e.g. in kind) to invest in the regional strategic planning effort. In addition to government and corporate investors, community foundations and other statewide or national foundations are likely resource controllers who might have stakes and investment interest in such regional strategic planning efforts; their participation might be either direct or indirect in nature. Members of the planning steering groups must be able to devote and commit significantly to this effort, such that there is effective participation and representation of the constituencies who have important stakes in this strategic planning project.

Each group of stakeholders can bring assets to the strategic planning effort. For example, representatives from tertiary education, including universities and community colleges or polytechnics, can contribute their research and/or training resources to advance the *awareness* stage and the ongoing surveying, monitoring and evaluation activities of the *talk* stage of the ALERT Model. These stages of the Model are elaborated on pp. 12–14 and pp. 131–3. Other representatives of the group will have additional capacities, social capital and linkages that can complement and be coordinated with the assets that the other members will bring to each group. Retirees from the business, corporate and institutional worlds can be especially valuable, in that they have the time and often the expertise to invest in the regional strategic planning process on an ongoing basis.

Especially during plan-implementation phases, it is critical to make assignments of actors in needed roles. For example, the government of the UK created the office of the e-Envoy in the Cabinet Office (Cabinet Office n.d.). The actor in this role is charged with facilitating the effective delivery of e-government services to its customers. Refer to Part V, section on Relational Planning Concepts, A–Z; see e-Envoy, p. 198. At the regional and local level, similar roles may be instituted to ensure that new technology-based functions have leadership and explicit attention.

Finally, it should be noted, by necessity, that these behaviors and principles suggested here are generic, not exhaustive and comprehensive. They represent initial directions that are intended to be helpful as the planners and leaders of a city-region grapple with taking action and begin to activate the *talk* component of the ALERT Model framework. The Model necessitates place-specific tailoring and application. For an example of a successful type of outcome that could function as a benchmark case, see pp. 78–9 on the Kitchener–Waterloo region of Canada.

ALERT Model

Background to the ALERT Model

It was not long ago that planning practitioners had few directly applicable theoretical resources or empirical findings to draw on to inform their practice as the new-economy global dynamics and new planning challenges emerged. That is the case no longer. There now are cases from which to derive lessons and there is a growing body of appropriate, rich and diverse planning theory that can offer general guidance for the practice of planning. Relational theory and its principles represent promising conceptual resources for the regional planner who is engaging the new dynamics and complexities of the global knowledge economy and network society. As portrayed here, a number of related conceptual constructs have been introduced or used to illustrate some ways that these concepts can be useful in contemporary regional and local relational planning practice.

Over the last several years, Corey and Wilson have been applying the Graham and Healey framework to the empirical context of the state of Michigan. Some of these applications have included several presentations that are available online (Smart Michigan n.d.). Wilson and Corey also have applied the principles underpinning the ALERT Model elaborated below to several cases in Asia – i.e. in Southeast Asia and to Korea at several scales or layers respectively (Corey December 18, 2003; 2004a; September 22, 2004c; Corey and Wilson 2005b). The next section summarizes the Model's elements, and illustrates their applications and integration with the Graham and Healey framework for relational planning practice.

Theory of the ALERT Model

As noted above in Part II, Graham and Healey's relational planning framework (1999) consists of four elements:

1 *Relations and processes*: this element has been operationalized by Corey and Wilson as the functions, factors and principal segments of the e-Business Spectrum (see pp. 45–51).

2 *Space and time*: in addition to the spatial organizational patterns of concentration and dispersion of the production, consumption and amenities functions, the space part of this element may be illustrated and operationalized by means of the six locational and economic development theories that have been reviewed by Plummer and Taylor (2001a; 2001b; 2003), and by means of a composite measurement construct that was based on an integration of these six theories, and has resulted in two key composite criteria that were derived as critical to inform future strategies – i.e. investment in human capital and the development of an enterprise culture. The time part of this Graham and Healey element was illustrated and operationalized by means of the Program Planning Model (PPM) (Van de Ven and Koenig, Jr Spring–Summer 1976). PPM is a planned-change process that proceeds through time; it incorporates relational propositions, especially by differentiating planning phases by role, actors and their principal contributions by stage or phase to the overall planned-change process. The importance of the "moment" in time and in the plan-implementation also was incorporated into the operationalization of the Graham and Healey framework (Sull July–August 1999; Sull 2003).

3 *Multiple layers of power geometries by place*: these have been depicted empirically as Michigan-specific layers and Asian-specific layers of mapped knowledge-economy functions and their respective spatial distributions, plus the two critical dimensions of mediation as represented by a juxtapositioning of areas of e-responsiveness with areas of economic distress. For purposes of illustration, these layers are exemplary of a hierarchy of units of planning that encompass the range of functions and creative tensions that typically exist as representative of various power geometries.

4 The *power of agency in negotiating between the layers of power geometries*: this element may be illustrated via discussion of the case of Michigan. Michigan regional planners contextualize and apply the notion of "agency" from the perspective of US Economic Development Administration (EDA) funds-seeking for projects of the state's regional Economic Development Districts. The planners accomplish this by making Comprehensive Economic Development Strategy (CEDS) plans that are submitted to the EDA. Successful CEDS plans are congruent with the employment enhancements that occur in the innovation and knowledge economy – e.g. higher wages – articulating the core values of the imperative to engage in inclusive development as a means of leaving no region and locality out of the planning in the new era of the global knowledge economy and network society.

This fourth relational planning-practice guiding element also included a discussion of planning scenarios that are intended to illustrate how the spatial unevenness of differential e-responsiveness and the varying locational distribution of economically distressed areas may serve as mediation dimensions in negotiating and prioritizing uneven power relations across Michigan and Southeast Asian regions. These presentations also have included additional planning scenario construction for Michigan with (a) the application of the product life-cycle theory; (b) suggesting human capital development and enterprise-culture links to the Michigan state government's-supported technology clusters, or Smart Zones; and (c) suggesting linkages of various layers of tertiary-educational institutions that make up a planning region's learning environment and as related to its technology clusters and distressed communities.

In sum, these four elements and their relationships, as articulated in 1999 by Stephen Graham and Patsy Healey, provided us with the theoretical inspiration for the construction and subsequent testing of the ALERT Model (see Figure 14). This figure is an illustration of our operationalization of Graham and Healey's relational theory-based framework for improving planning practice for today and tomorrow.

The Model

The European urban studies and urban planning scholars referenced in Part II have inspired and influenced the construction of the ALERT Model. Conceptually, the development of the model represents a transition from the urban studies of Europeans Jean Gottmann, Stephen Graham, Patsy Healey and others into urban and regional planning research of the complex relations among today's economic and societal dynamics to acting on some of those findings by tailoring and inventing one's practice of relational planning to the concrete realities of one's local and regional locations. Among the many lessons from these studies, we have been influenced particularly by the critical need to change the conventional planning mindset and the critical need for experimenting with new-governance planning behaviors.

This section introduces the ALERT Model as an organizing framework for the practice of contemporary development planning in this era of the growing pervasiveness of the global knowledge economy and network society. When all five elements or phases of the Model are working together, ALERT stands for the need to have improved planning practice that takes into account and is inspired by the principles inherent in Graham and Healey's relational planning framework.

The ALERT Model's five elements are intended to: instill (1) *awareness* among an area's stakeholders in order ultimately to change their planning mindset and behaviors to reflect today's new economic and social realities – i.e. to behave from the knowledge that localities and regions are integral parts of (2) *layers* of various economic-social-cultural functions at different scales

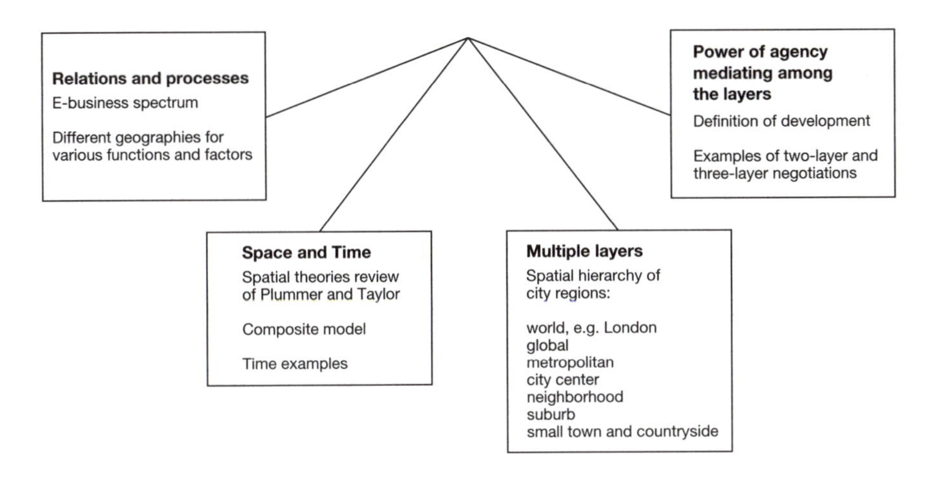

Figure 14 *Theory of ALERT Model*
Source: Authors (inspired by Graham and Healey, 1999).

as influenced by diverse forces and levels of globalization. The ALERT Model uses the electronic business spectrum, or for short, the (3) *e-business* spectrum's economic functions as a way to organize the principal activities of the global knowledge economy, locally and regionally. The Model calls attention to the need for the planning to be (4) *responsive* to the opportunities that can be identified by researching the special niches, advantages, differentials and complementarities that each locality offers for its potential development in the future. The ALERT Model requires operating under new governance and collaborative relationships among key representatives of the area's principal stakeholders; in order to initiate and sustain this new governance and new planning behavior, these development actors engaged in the ALERT process must relate to and (5) *talk* with each other on a sustained basis. In this case, "talk" is a metaphor for this new form of cross-institutional and multiple-organizational behavior that must be continuous over a long time-frame and closely coordinated. Consequently, one should be clear that this fifth element of the ALERT Model involves "talking," but it also involves continuous interactions among the region's representative stakeholder institutions, businesses and individuals. Given the need for ongoing planning and plan-implementation, the *talk* element of the model should be an activity measured in years and decades, if the transformation of local human capital development and enterprise culture is to be fully imprinted on a region's knowledge economy development planning. The five elements of the ALERT Model may be conceived and operationalized as a process of interdependent stages that, over time, need not necessarily be planned and executed in a linear fashion. Indeed, the relationalities inherent in the Model's complete realization necessitate the routine practice of relational thinking and behaving.

Awareness

The goal of the *awareness* phase of the ALERT Model is to learn, and in the process, be motivated sufficiently for the community's stakeholder representatives and their planners to commit to a sustained process of fundamental long-term planning-behavioral change. In this context, the intent is to be aware, to be alert and to have knowledge that enables informed action to be planned and taken. Awareness here also means that inferences can be drawn from knowing. It is appropriate in the era of the knowledge economy that the strategies and plans for our communities should be based on knowledge of the condition, assets and resources regionally and locally – and within the context of the global forces and opportunities of our localities. The outcome of awareness is "actionable knowledge"; refer to Part V, section on Relational Planning Concepts, A–Z, p. 189.

The criticality of the *awareness* phase of the Model may be illustrated by the following event. In spring 2005, we conducted a simple survey of a small group of representative community stakeholders from a county in Michigan. They were part of a group who had volunteered to be part of a "citizens planners" program. The topic of that meeting was "An Assessment of the State and County's Knowledge Economy." Before the session began, we asked the participants to record their perceptions of the most pressing planning issues in their region today. One result was particularly noteworthy, especially given the topic of the meeting. Of the twenty citizen planners, only one participant listed education as one of the most pressing problems facing the region. By the end of the meeting, they documented that they were much more sensitized and *aware* of the criticality of education and human capital development, as well as a wide range of other functions related to regional and local development and competitiveness in today's global knowledge economy and network society.

Human nature being what it is, there is a strong proclivity to be heavily engaged in the routine activities of the day-to-day, and in the process of the every day we often, indeed too often, miss out on the larger forces and issues that also affect our lives. Organization development expert Warren Bennis has labeled this phenomenon the "unconscious conspiracy" (Bennis 1989: 14–15). A core contribution of effective knowledge-economy era planners is to ensure that the development leadership of the region does not succumb to the unconscious conspiracy and avoids being Rip Van Winkles. In a commencement address in 1965, Martin Luther King, Jr used the story of Rip Van Winkle to remind his audience that Van Winkle slept through a revolution. When Van Winkle went up the mountain, the image of King George III of England was on the sign of the inn in town; when Van Winkle returned from his long sleep, President George Washington's image then was on the sign. The moral of the story for Dr King was,

> there are too many people who, in some great period of social change, fail to achieve the new mental outlooks that the new situation demands.

> There is nothing more tragic than to sleep through a revolution. There can be no gainsaying of the fact that a great revolution is taking place in our world today.
>
> (King June 1965: 2)

What was true for Dr. King in the 1960s is true for us also in the twenty-first century. Only our revolution is technology induced and has been transforming our economies and our societies. Similar to Rip Van Winkle, in general, we have been sleeping through this technology-economic revolution. We cannot afford this. Our communities cannot afford this. If we wake up and become alert, then what should we do?

An initial activity of awareness is to frame the planning activities (Lakoff 2004). Framing provides the context for developing and designing the strategic planning approach to be followed by the planners and stakeholder-leaders of a region and its communities. Framing in this context means taking a fresh strategic view of one's community; it involves knowing (1) the facts of the community's technology assets, (2) its economic position and (3) the special qualities of the community. These three sets of knowledge are the basic building-blocks for new strategic planning in today's new era of the global knowledge economy. How might your region and community frame and approach its needed new planning? The ALERT Model offers an answer to that question.

Discovering the region's relative position amid the many economic and social changes affecting the locality is an important starting point for initiating the application of the ALERT Model. Working with the representatives of the principal stakeholder groups of the region and its localities, the relational planner, and her/his constituents and clients need to develop a working understanding of and commitment to the need – indeed, the urgent need – to change from the old ways of doing planning and formulating policies. This needs to be achieved by educating the planners and stakeholders that technology and knowledge are transforming wealth-creating work from physically based functions to knowledge-based functions. This awareness, in turn, should translate into planning and policy action by means of continuous recognition that highly influential economic forces today are global, and they rely increasingly on knowledge and technology-enabled use of knowledge as factors of production and wealth creation, in addition to the traditional factors of labor and capital. We have applied and tested the awareness building and education processes initially by mapping the spatial distribution of ICT infrastructure and access to that infrastructure, especially to the Internet via high-speed broadband access. In addition, we have plotted the changes of these and other information and telecommunications technologies over time; these have ranged from landline telephones, to cellular phones, to personal computers and Internet changes. Empirically, these mappings and plottings have been conducted to date for Michigan and the countries of the Southeast Asian region (Corey and Wilson 2005b). At a minimum, an outcome of the initial awareness-creating activity is a single spatial and temporal layer of understanding, such as maps of local and regional conditions that depict the relative current status of technology and economic conditions. Because of the understandings developed during the awareness stage, layers of specific operational regional or local awareness can be achieved.

From the perspective of regions and localities around the globe, much more attention needs to be paid to, and concern for international competition and the approaches that are used by global competitors in other regions. It is important to understand the success factors that competitors have used. As noted earlier, the role of benchmarking and the studying of best practices can be helpful in systematically gauging the competitive environments external to the locality and its planning region.

This phase of the Model needs to have resulted in widespread familiarity with the technology-induced changes, trends and issues facing the region and its localities (refer to Part I, p. 14). To be effective for the intelligent development of the community, the Model must function on a continuous ongoing basis. Among other outcomes, this means that there needs to be an institutionalization of feedback from the community and continuing lifelong education; for example, planners need to create learning organizations to maintain awareness. Refer to Part V, see section on Relational Planning Concepts, A–Z, Learning Regions and Innovative Milieux, pp. 208–9. For more detail on the critical planning practice activities of this phase of the ALERT Model, refer to Part V, see section on *awareness* phase, pp. 228–9.

Layers

The initial layers of awareness described in the section above can serve as the basis for generating additional layers of substantive understanding for a particular place. By means of these multiple layers of understanding developed during this, the "L" or *layers* stage of the ALERT Model process, the purpose of this phase is to produce a deeper awareness of the many and different geographies and temporal relations and processes that are required to track some of the region's major technology and economic relations, and then to engage in successful relational planning. For example, in the Michigan case, five layers were developed (see Figure 15). For the Southeast Asian case discussed below, three layers of strategies were developed. Michigan's layers, for the purposes of analyzing the state's knowledge economy, consist of: (1) the two world knowledge competitive regions of Grand Rapids and Detroit (Huggins *et al.* 2004); (2) biotechnology and biosciences firms mostly clustered in the Michigan Life-Sciences Corridor with its four principal research institutions and universities; (3) the technology clusters, especially as represented by the Michigan Smart Zones, noted below; (4) universities and community colleges that are widely distributed throughout the regions of the state; and (5) the extensive empty areas that are largely undeveloped and represent important natural resource-

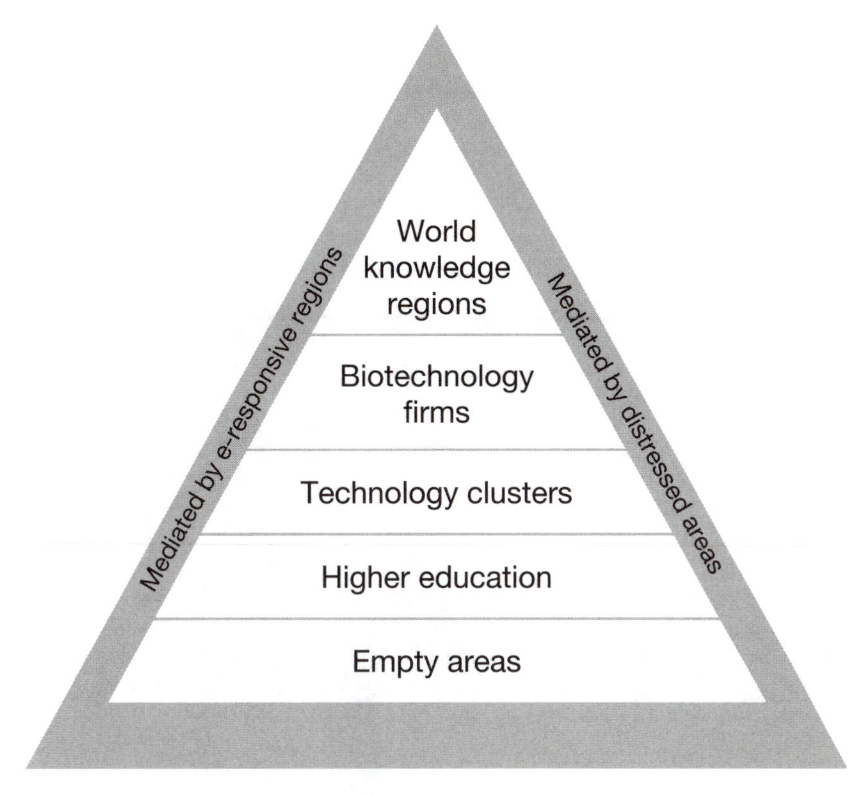

Figure 15 *Michigan Layers*
Source: Authors.

based amenities for enhanced quality of life. These multiple layers serve to instill a more operational, but still highly partial understanding of the economic and social relationalities, complexities and their many dynamics as represented by means of the multiple layers.

In order to understand the distribution and current condition of some of the key regional and local assets in the Michigan case, selected layers representations are offered. These are layers of critical content. They are simple substantive measures that are surrogates for the complex relationalities that organize into the many and highly variable structures and flows of today's regional and local economies and societies. These illustrative planners' layers basically function within the planning-administration geographies of the state – i.e. its planning and development regions (see Figure 16), counties, and local cities and communities. Using the kind of information that is readily available to local planners, for these planning regions and the counties that comprise these regions, a selection of relationships was mapped in various combinations or overlays and scales in order to illustrate simple spatial relationality.

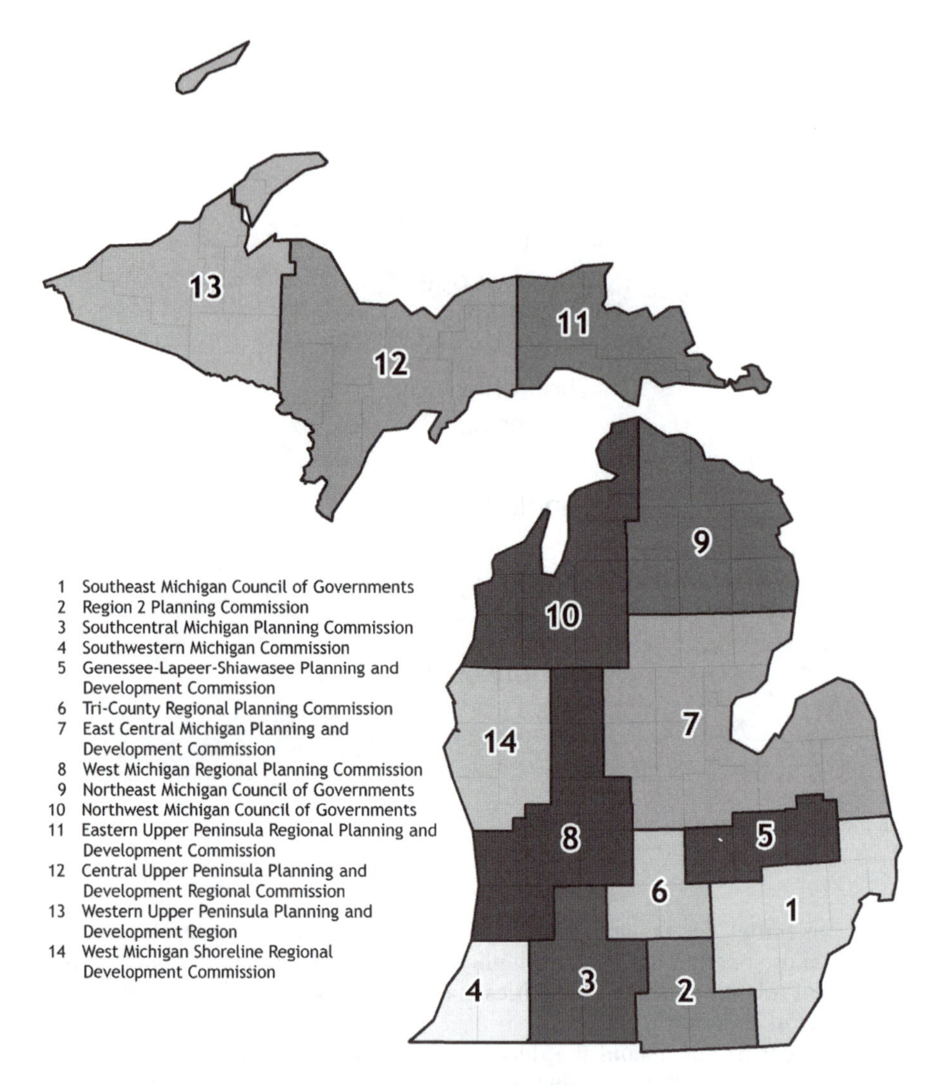

Figure 16 *Michigan's Planning and Development Regions*
Source: Authors.

e-Responsiveness This is a two-layer relationship. This layering consists of (1) counties by the effectiveness of their economic development websites and by (2) planning region, composed of counties, with varying degrees of readiness by their regional planners to plan for the engagement of their localities with the global knowledge economy and network society, including digital development. This latter layer is a spatial measure of localities' responsiveness to compete in today's e-business environment.

Priority planning for digital development This is a two-layer relationship. It demonstrates the operationalization of how regions and localities might begin to address the value of equity and access for all the state's regional economies and societies. The following layers were mapped. (1) By county and at the state level, all economically distressed communities were mapped. (2) This distribution was overlain with a map of those two-geographies e-responsive counties, just noted above, that were surveyed and found to be responsive, somewhat responsive, and less responsive to planning for engaging the global knowledge economy in their economic development planning. This two-layer combination could form the basis for prioritizing planned interventions to ensure that the poorer served and less high-speed Internet-accessible localities of the state and its regions also might compete for their development in the global knowledge economy and network society (see Figure 17). For the obvious four planning-scenario priorities in this case, see p. 64 in Part II.

Product life-cycle planning This is a three-layer relationship. It was inspired by the product life-cycle theory, which is described in Part V below, see section on Relational Planning Concepts A–Z, Product Life-Cycle Model, pp. 213–14. A hypothetical relational spatial planning process and example might develop as follows (see Figure 18). In the first, the formative stage, a life-science innovation may be characterized by new knowledge being generated by a basic researcher or research team in a laboratory on the campus at Michigan State University (MSU) in East Lansing. After appropriate gestation time for this basic research project, the findings have progressed to the next, the second stage of development – i.e. the *intermediate or growth stage*, such that the life science innovation has demonstrated a proof of concept, and this has been accomplished in a larger laboratory off the MSU campus at the nearby University Corporate Research Park in Lansing. The life-science innovation has matured to the third stage in its life-cycle – i.e. the mature stage, such that the Dow Chemical Company has invested in the innovative product and has taken the demonstrated proof of concept research to its Midland, Michigan, manufacturing location to move the product into production. Over time, as the production of the life-science product declines to the stage – i.e. the declining and/or congestion stage, where it assumes relatively low innovative status, it lends itself to being hived off to more low-cost peripheral locations, such as to Michigan's Upper Peninsula, for example, or out of state or out of the US to lower cost locations in the global economy. Another characteristic of the fourth stage of the product life-cycle is the tendency for localized congestion to occur as other production companies with related content production locate in a cluster for this product. With such situations, there also is a tendency for related innovations to be spun off from the production cluster. This requires new space for this new idea (Swann 1999; Audretsch and Feldman 1996). This might mean locating the research effort for the new spin-off at the MSU laboratory campus environment, or at a totally new location such as at a research facility in Traverse City, Michigan, that has the

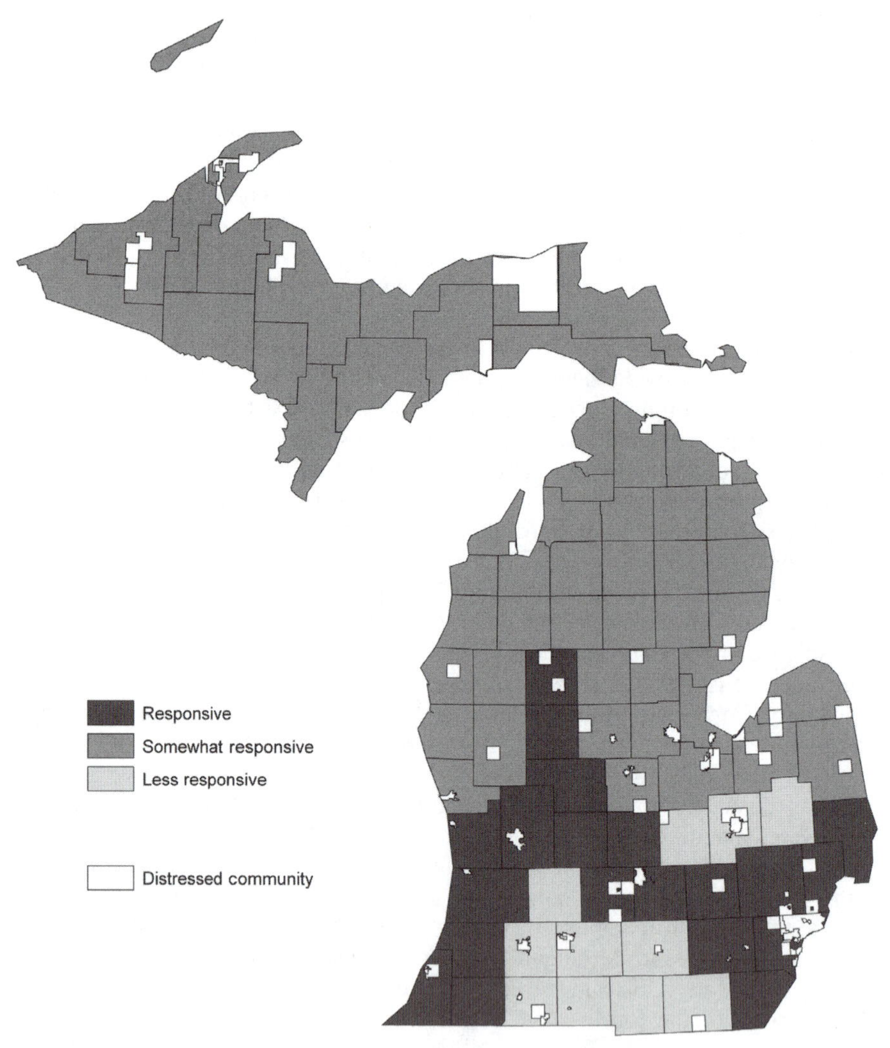

Figure 17 *Priority Planning for Digital Development in Michigan*
Source: Authors.

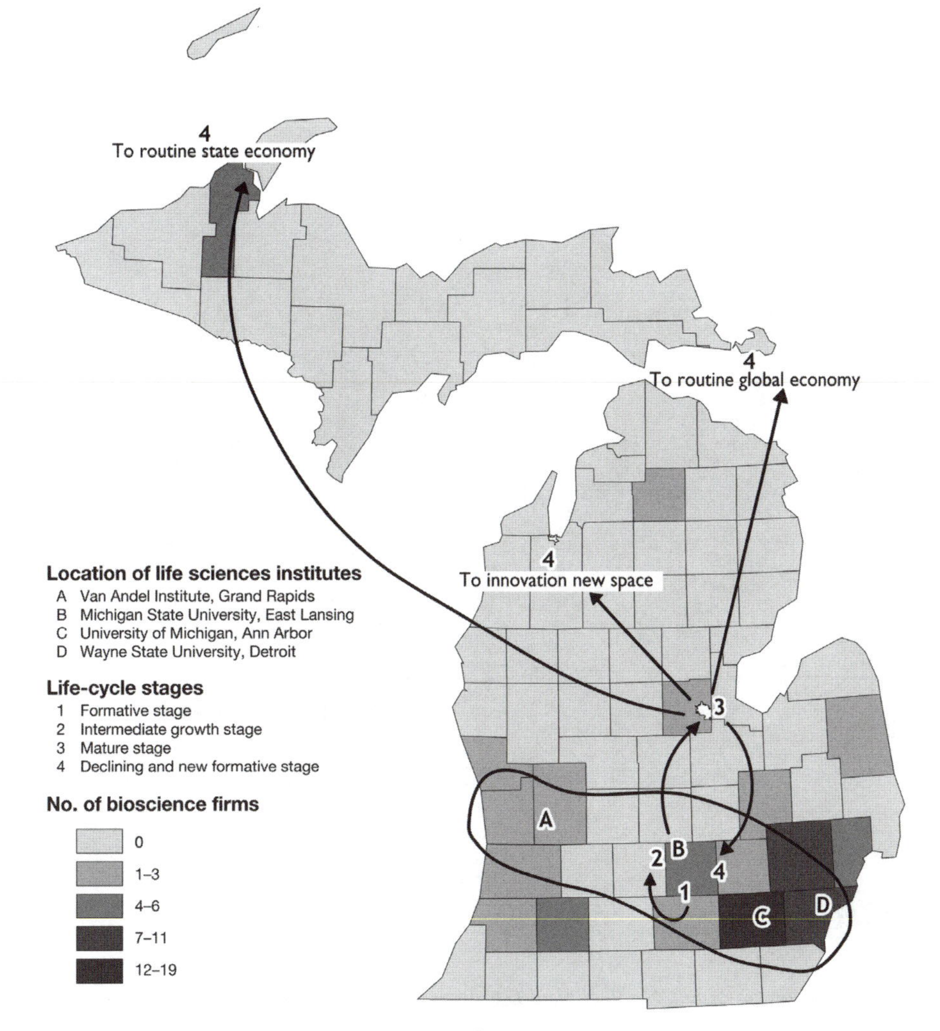

Figure 18 *Product Life-cycle Planning in Michigan*

Source: Authors.

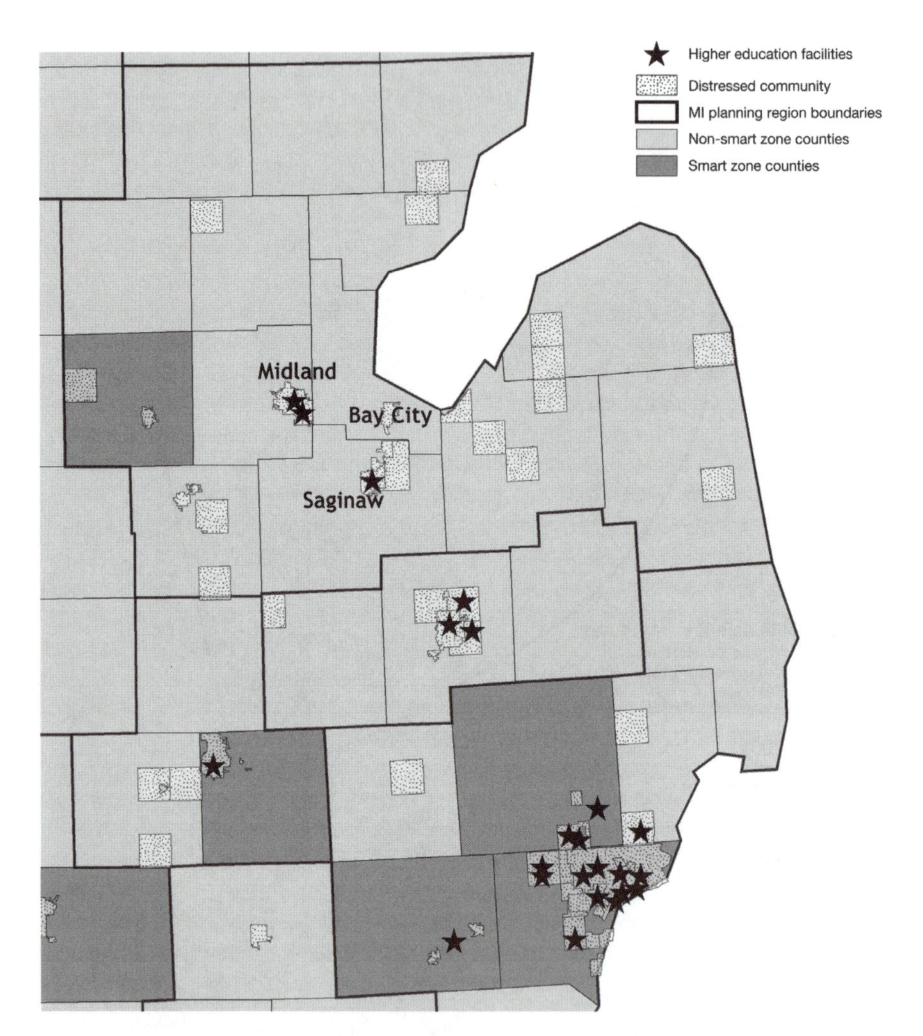

Figure 19 *Toward Equitable Intelligent Development Planning in Southeast Michigan*
Source: Authors.

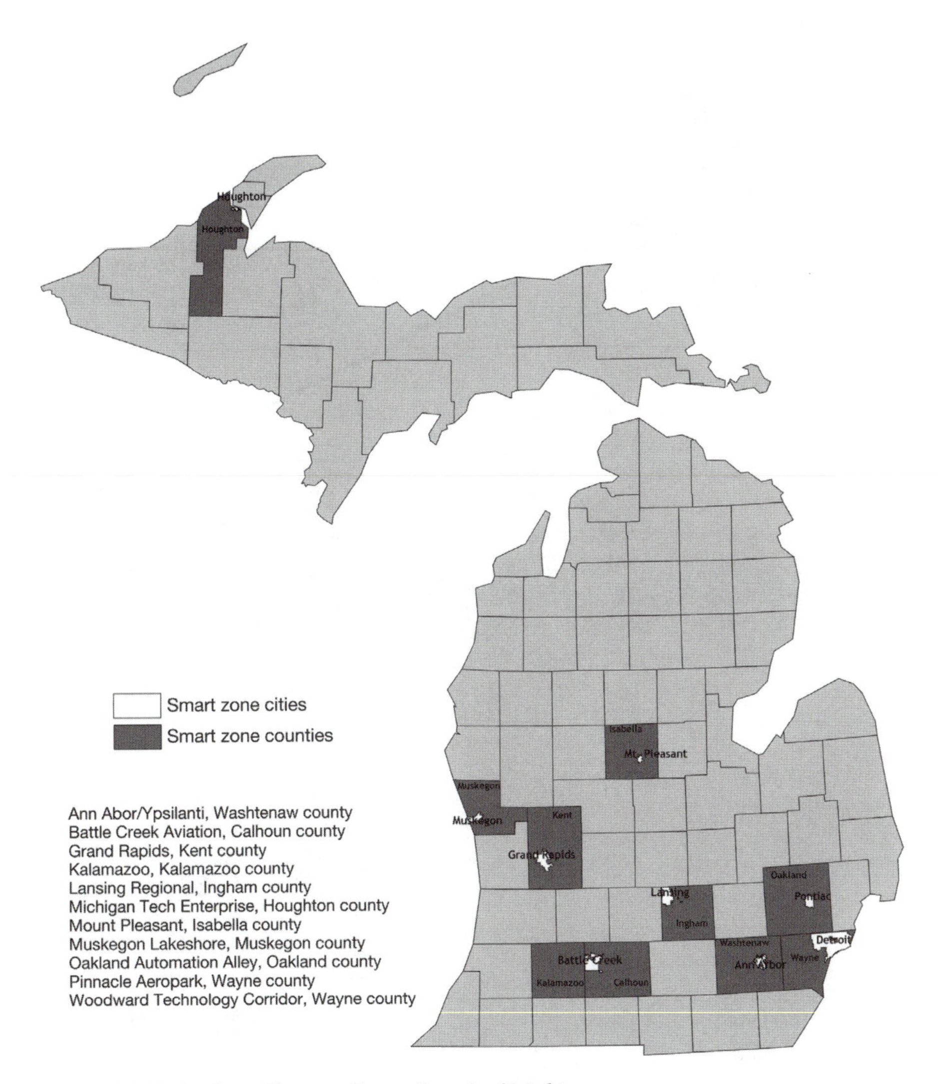

Figure 20 *Technology Clusters (Smart Zones) of Michigan*

Source: Authors.

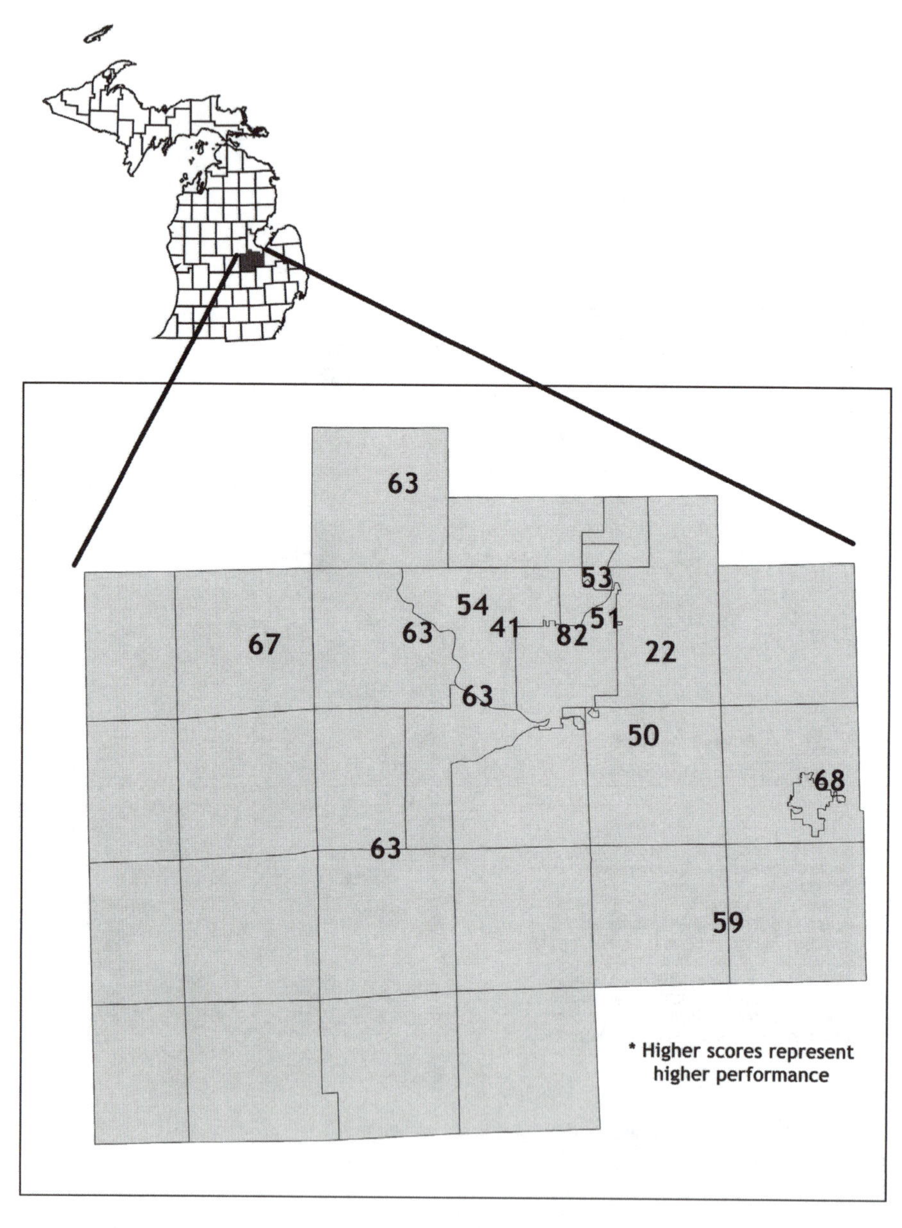

Figure 21 *Distribution of High School Performance Scores in Saginaw County, Michigan*
Source: Authors.

appropriate research talent, facilities and venture capital environment to make that fresh location attractive.

Figure 18 shows that the illustrative layers, in sum, were: (1) Michigan's Life Sciences Corridor was represented on the state map (as the line encircling the A–D life-sciences research institutes), and (2) by the four developmental stages of the theory, a scenario was constructed and mapped to depict how a hypothetical research and development-based life-science innovation might realistically (3) flow from research institutes and universities, and evolve to compatible host commercial corporate locations. Some of these locations had nearby tertiary educational training institutions to assist in the spread of these intellectual property benefits and spillovers, and some did not have such institutions in close proximity. Intentional and planned partnering, twinning and joint relationships among science and technology actors and agencies enable these kinds of functional, developmental and spatial relations. They are natural to, and congruent with the product life-cycle of idea-to-production processes over time and space. Indeed, this theory-inspired hypothetical is illustrative of an "intelligent" development-planning scenario. These are the kinds of relational and spatial planning and layering concepts that can be utilized by regional planners in general, and Michigan regional planners in particular to be more effective practitioners in the new environment of the global knowledge economy.

Toward equitable intelligent development planning Initially, this is a three-layer relationship at the scale of several adjacent planning regions (see Figure 19). This set of relationships is composed of: (1) counties within which the operationalization of formal components of the local enterprise culture were mapped. These were technology clusters designated as Michigan's Smart Zones – i.e. areas within select cities that are intended to focus local technology-facilitated investment and economic development (see Figure 20). (2) Locations of universities, community colleges and other tertiary educational institutions, and (3) locations of economically distressed communities (see Figure 17). These three layers also were analyzed in conjunction with a fourth layer, (see Figure 21). This is a representation, at the county scale, of the locations of the secondary-level high schools and by the performance scores of their students on annual standardized tests. These four layers suggest the kinds of simple relationalities that, when analyzed holistically, can suggest actions, plans and priorities to enhance the human capital development of the region and its localities.

Analogous and similar mapping of layers also have been constructed for Southeast Asia. These figures demonstrate the generic use of the ALERT Model (Corey and Wilson 2005b). See Figure 22 and Figure 23 in mapped form. In this, the *layers* stage of the ALERT Model, emphasis is on geographic and temporal layers. The context of globalization and spatial interaction – i.e. the forces of external change and potential – need to be taken into account in strategic planning by a local region's planners and stakeholders. For example, digital and intelligent development strategies by Southeast Asia's most advanced

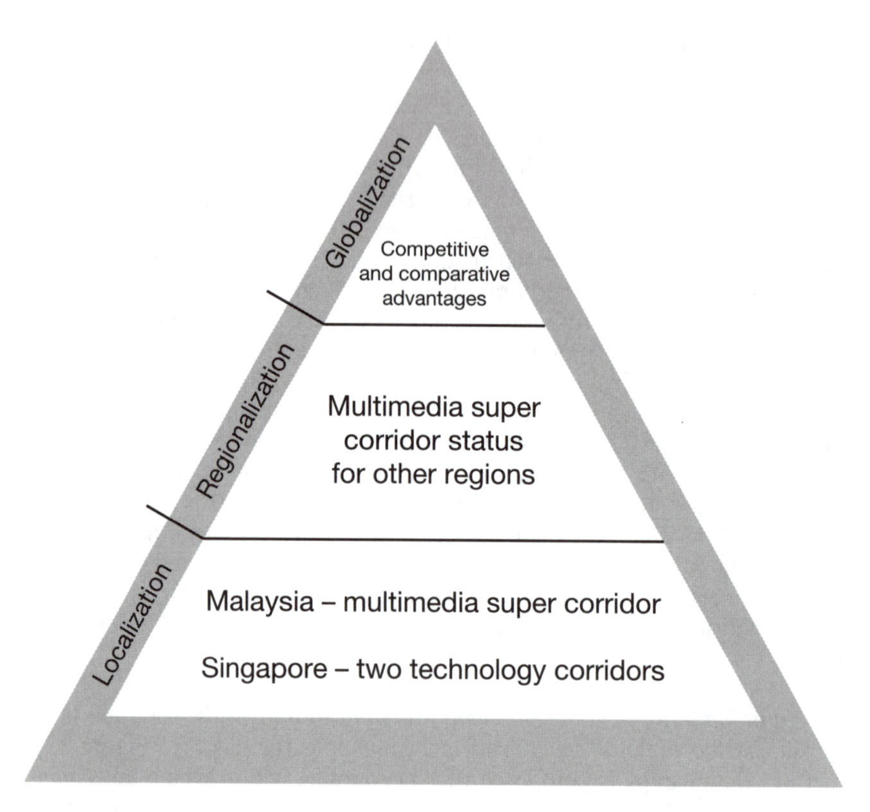

Figure 22 *Southeast Asian Strategies Layers*
Source: Authors.

global knowledge economies, Singapore and Malaysia, would plan localization strategies that capitalize on their existing corridors' plans and investments (Corey 2000; 2004a). Malaysia needs to spread the development benefits and incentives to other regions of its economy, beyond the Multimedia Super Corridor (MSC), which is represented on Figure 23 by the Kuala Lumpur International Airport, located at the southern end of the corridor. Consequently, a regionalization strategy is required for more equitable and even development from the spillovers and benefits of MSC status. Both Singapore and Malaysia's economies are highly interdependent with the Asian marketplace beyond Southeast Asia and with the global marketplace. This too requires globalization strategic plans that exploit their respective competitive and comparative advantages. Such explicit layered strategic planning emphases feed into and sustain awareness. Additionally, the internal layer of each national spatial organization demands planning attention, such that no area goes unattended in the strategizing. Temporal layers also drive the planning; organizing the strategies with explicit futures attention in mind is essential. Short-term, medium-term and

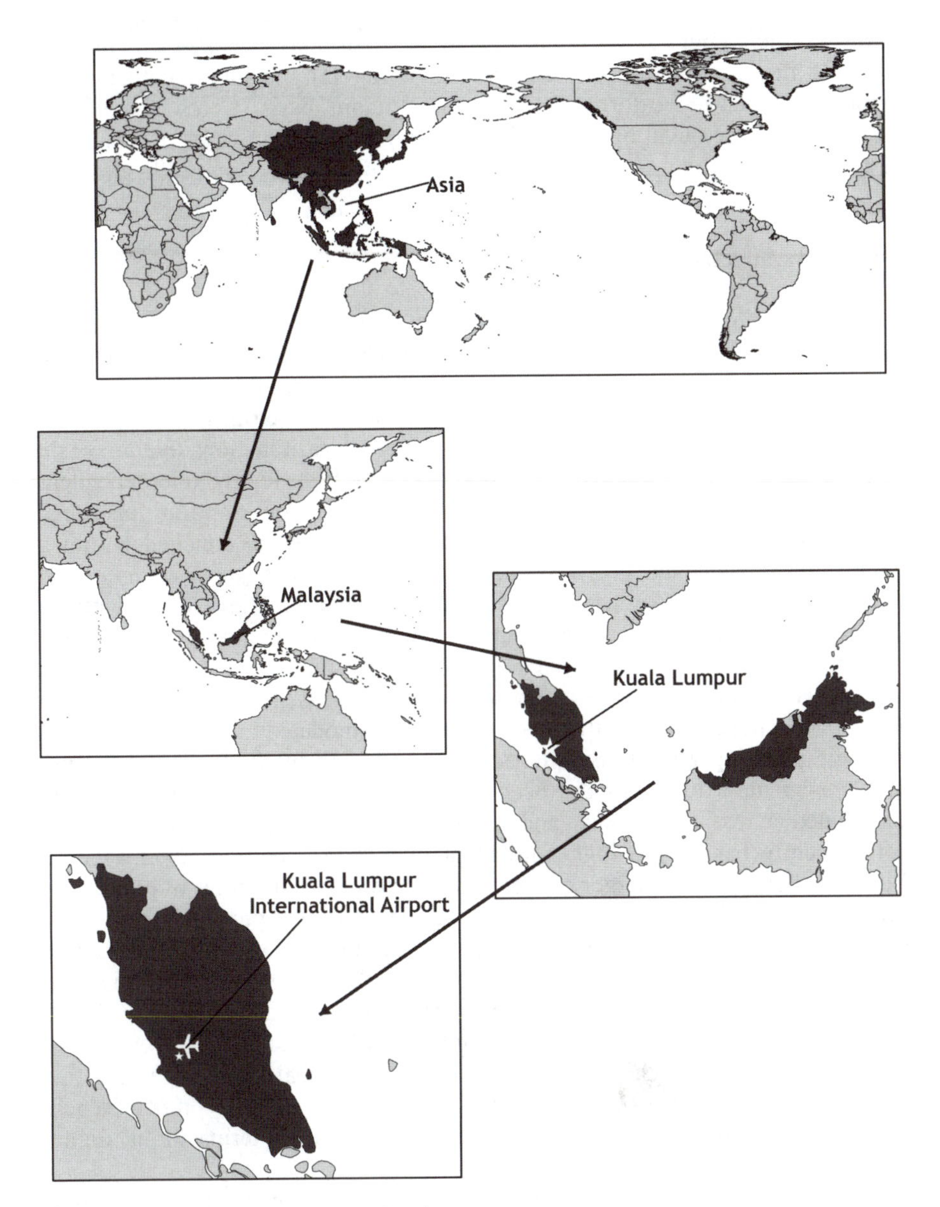

Figure 23 *Southeast Asian Layers (Map)*
Source: Authors.

longer-term future programs and actions must be planned. As the ICT infrastructure becomes more widely distributed and digital development matures through time, progress toward development goals may be assessed, thereby enabling future trajectories of strategic planning for the intelligent development that needs to be formulated on a continuous basis (Corey and Wilson 2005b). For more detail on the critical planning practice activities of this phase of the ALERT Model, see Part V, section on *layers* phase, pp. 229–30.

e-Business Spectrum

In the *e-business* stage of the ALERT Model, a more complete awareness may be developed by overlaying and explicating each component of the electronic business spectrum (see Figure 13 on p. 47; see also pp. 45–51 for an elaboration of the conceptual construction of the e-business spectrum). In a simplified form, the principal categories of economic and societal functions that are basic to successful and competitive development in today's highly interdependent economic world are: (1) production functions; (2) consumption functions; and (3) amenity and quality-of-life factors. These may be couched within a context of ICT-enabled or "e-business" terms to be congruent with today's new technology and knowledge-based development realities and potentials. These functions are the primary content or substantive economic and societal development activities that are facilitated by the Internet and especially by high-speed broadband networks. These various functions have different geographies or locational patterns that need to be accommodated in the visioning and strategic planning for future digital development and intelligent development. While e-business functions are becoming more pervasive, they are unevenly distributed across the regions and localities of the world. Generally, digital infrastructure has been a highly urban phenomenon – i.e. the better accessibility is where there are larger concentrations and higher densities of population and enterprises. Rural communities and the most sparsely populated areas tend to be the least served electronically. Periodically updated digital infrastructure maps are needed to determine the location of those areas that are unserved and under-served by digital technologies. These patterns of unevenness should be addressed over time by regularly updated ICT plans so as to enable all localities and areas of the region to engage in the global knowledge economy and network society. In order to enhance a nation's overall competitiveness and quality of life, no region should be left behind in terms of its digital development.

This, the *e-business* stage of the ALERT Model, should involve considerable in-depth action research (Part V, section on Relational Planning Concepts, A–Z; see Action Research, p. 189) and action planning among the locality's principal stakeholders. Their research should involve careful and thoughtful engagement with the existing and potential development opportunities for the local area's knowledge-economy functions and factors. By these means, a fuller development of a collective awareness in, and ownership of new innovative strategies may be attained. This work should seek to identify the unique competitive, special qualities, comparative advantageous relations, the disparities, the differentials and the complementarities of the region and its communities. By organizing this further awareness-building according to the driving and instrumental functions of knowledge-economy production, consumption, and amenity and quality of life factors, an intentional positioning will have occurred and a platform been created for moving into a planning and action-taking mode within a policy context that will have been formulated collectively across stakeholder groupings. In sum, this stage uses the e-business spectrum to study and analyze – i.e. to engage in action research; it also uses the spectrum to begin to frame the strategic planning, including its content, and locational and spatial dimensions. The four layers mapping examples described above are illustrative of the initial kinds of e-business spectrum-stimulated scenarios that might form the basis for more elaborated new action plans, programs and policies.

The emphasis in this phase of ALERT modeling is based on planning discussions of how ICTs are changing and should change the content inherent across the entire spectrum of the region's business functions, and how the planning area's places may affect, apply and expect specific benefits and costs of ICT use. This is the phase of the Model that shifts the strategic emphasis to intelligent development, based on the presumed momentum of the continuous digital development rollout that will have been set in motion for serving the entirety of the region and its localities. For example, see Figure 24; this is the e-business spectrum introduced and described in Part II (see pp. 45–51). This version of the e-business spectrum includes a selection of policies of State

Digital development and intelligent economic development: the goals are to continue ICT infrastructure rollout and to shift emphasis to creative competitive content		
Production functions	*Consumption (e-commerce) functions*	*Amenity and quality-of-life factors*
Science and technology-driven research and development (C) – S2 billion jobs tomorrow bond proposal	Online procurement: B2B and B2G and G2G (D) – Michigan Deal – bulk puchases	Innovative social, cultural and institutional activities (C and D) – cool cities initiative
Commercialization of products and services (C) – Technology tri-corridor	Online retailing: B2C and G2C (D) – permitting process	Natural environmental attributes (C) – Clean Michigan bond initiative
Business and producer services (C) and manufactured products (D) Public and government producer services C and D) – e.g. regulations, taxes, info., etc. – Smart Zones and MBDA	Value-added complementarities between electronic (clicks) and physical (bricks) channels (C and D) – e.g. licence plate registration renewal	High-quality education and human capital capacity-building and talent development (C and D) – Jobs today initiative

Figure 24 *e-Business Spectrum (Michigan Policies)*
Source: Authors.

Planning for your region and its localities to do the "business" of electronic-driven economic development		
Production factors	*Consumption (e-commerce) functions*	*Amenity and quality-of-life factors*

Figure 25 *e-Business Spectrum (Blank Template)*
Source: Authors.

Government; this illustrates the kind of policies and programs that need to be planned to move the region's content-based intelligent development planning to action. Also, see Figure 25; this is a simple template to illustrate that the local planners and their client stakeholders need to develop systematically the regional and local policies and programs required to enhance the area's competitiveness in the context of the global knowledge economy. For more detail on the critical planning practice activities of this phase of the ALERT Model, refer to Part V, see section on *e-business* phase, pp. 230–1.

Responsiveness

As a preparation for being responsive, the e-business spectrum may be used as a kind of preliminary template at regional and local scales to plan for the principal functions, factors and relations that are critical and strategic for engaging today's and tomorrow's transforming global political economy. The *e-business* stage of the ALERT Model is based on collective and action research among the planners and the stakeholder-representatives of the region and its localities. With the pre-planning research of the *e-business* stage having been organized by today's basic economic, societal and cultural functions of the knowledge economy, the community then is positioned to be opportunistic to advantageous relations, potentialities and disparities of the place as revealed by the local analyses as stimulated by the template of the *e-business* stage – i.e. see Figure 25. The more responsive the place is to the development opportunities, inequalities and relations of the global knowledge economy, the more likely it will be a successful regional, national and international competitor. Disparities, such as digital divides of different degrees of access and ability to use high-speed broadband connectivity for example, is an issue that needs to be addressed if the entire community is to be served equitably and fully capacitated to compete (Servon 2002; cf. Schön *et al.* 1999).

In the US, large telephone companies are lobbying legislators for laws prohibiting local governments from subsidizing high-speed Internet connections. With the US behind many countries in per capita broadband usage, with many communities and segments of the population without such service, numerous local governments are struggling to ensure that all segments of their population and all areas of their localities are provided with affordable wireless broadband access. For example, the city of Philadelphia, Pennsylvania, "will become North America's first municipality to offer affordable broadband Internet access as a public service to all residents" (Morrison 2005). The city's wireless initiative has drawn active criticism from firms that are Internet providers. They say that the building and operating of such ICT infrastructure and services is best done by experienced private-sector companies and not by a municipality with no relevant experience and with public monies. City leaders say that market forces will leave low-income communities unserved. Oakland county, Michigan, in the Detroit metropolitan region also seeks to provide broadband access throughout its extensive area with its "Wireless Oakland" initiative. However, the county will rely on the private-sector companies to rollout wireless broadband service. Under this partnership with the county, companies can use their communication towers and the thousands of other physical assets. In turn, the service providers will offer lower bandwidths of Internet service free of charge which is intended to enable low-income residents to obtain ready access to the Internet. In return, the private-sector service provider partner is authorized to charge a fee for the higher bandwidths that are faster and enable more applications (Oakland County, Michigan n.d.). These two approaches demonstrate different paths to the same goal and they illustrate how important some local communities prioritize universal access to broadband and in the process enhance their competitiveness and attractiveness for the full spectrum of e-business and investment. The large telephone companies say that places without broadband should offer those companies tax incentives in order to obtain service (Drucker and Li June 23, 2005).

In order to move intelligent development visions and plans to action and the realization of actual development opportunities, effective partnerships and working relationships must be forged throughout the region and its localities. Here is where social capital investment can produce important returns in the context of ICT-enabled development (Huysman and Wulf 2004). Depending on the particular circumstances locally and regionally, such relationships need to span economic sectors, institutions and individuals. In many places around the world, for effective strategic planning to be moved to plan implementation, new forms of governance must be devised, practiced and perfected. This will take time and maturation. It also will require leadership, champions and parallel divisions of labor. The results and timing of these actions require well-coordinated interdependence and complementarity. As strategic planning outcomes occur however, they can serve increasingly as partial foundations from which to build and extend follow-on planned actions – that is, the more pieces of the overall development strategy that get put in place, the clearer becomes the next set of actions that need to be planned and taken. In order to be responsive, therefore, attention must be paid to, and actions taken to address the unevenness of the distribution of electronic infrastructure and e-business development opportunities that were noted on p. 64 regarding the *layers* and *e-business* phases. Without such attention, inequities and economically distressed areas might become even more distressed and entrenched in the new knowledge economy and network society of the sub-national region and its localities.

Having studied and analyzed the region and locality's conditions and their relative positions, the planner and the region's stakeholder representatives are prepared to begin to influence future development. Critical, and at the core of this activity is to move the community from a stance of reacting to these assessments to one that is proactive, entrepreneurial and increasingly is prepared, and thereby quick to respond to opportunity.

It is in this context that intelligent development should be pursued and practiced. Premised on the opportunities and niche potentials identified in the *awareness* stage, and drawing on the evolving body of appropriate theory, regional and local planners need to lead the strategy development and the plan formulation. Scenario planning, in particular, should be conducted as a useful way to explore and exhaust strategic development options, and to generate widespread community ownership in well-considered and planned responses for local opportunities to be derived from the global knowledge economy and network society.

It is important to monitor progress on the fundamental strategic elements noted in Part II (see pp. 44–67). In order to be more fully and continuously aware, and thereby be positioned to be responsive and opportunistic, regular community and planner surveys are essential. The following five elements, among others, should be periodically assessed. One of the objectives for the survey is both to derive information (including from the planners themselves) for supporting regional and local planners in their planning practice within the context of the global knowledge economy and network society, and to build on and test some of the key planning practice elements. Five critical elements are: (1) the development of *human capital* throughout the human life-span (i.e. from early childhood development through primary, secondary and tertiary formal education and in-service job development education and training); (2) achievement of an *enterprise culture* in the region that hosts and supports entrepreneurship and innovation by means of cross-sectoral collaboration and partnerships among the business and nonprofit institutional communities; (3) the need to change the *mindset* of regional and local planners so that their practice behavior is congruent with the complexities and uncertainties of the global knowledge economy; (4) an enterprise culture can be created and sustained by means of new *governance* behaviors that transcend the major stakeholder interests of the region and seek to ameliorate the fragmentation of the various governments as together they seek to converge planned actions at the local level; and (5) explicit and activist attention to the region's *economically distressed communities*; this means identifying technology and economic disparities, their spatial distribution and developing strategies and plans to attain more equitable distributions of the local opportunities and potentialities from the global knowledge economy. Admittedly, these elements are diverse, but they represent needed outcomes and critical planning practices that are prerequisites to success within the context of the global knowledge economy and network society. Consequently, it is helpful to the region's overall development-goal attainment to obtain empirical and perceptual information that enables local planners regularly to compare and contrast this five-element assessment framework to current planning practice and what the planners perceive their practice needs ought to be. Thereby, continuous fine-tuning and updating of planning approaches are built into the Model, making it routine that the new planning behaviors might be corrected to ensure that they are responsive to the constantly changing conditions and their relations

In sum, the responsiveness stage of the ALERT Model recognizes that there are different types and levels of response throughout the region. As a consequence, the strategic planning groups must proceed to discuss and plan what it will take to be more fully and equitably responsive. In addition to understanding how responsive places have been, actions to address fuller responsiveness require leadership. This means assignments and designations of champions and support groups whose task is to ensure that the solutions planned for improved local responsiveness are implemented, monitored for progress and evaluated for conformance to the goals of the responsiveness strategy. For more detail on the critical planning practice activities of this phase of the ALERT Model, refer to Part V; see section on *responsiveness* pp. 231–3.

Talk

While this stage of the ALERT Model includes verbal discourse, the word "talk" here is used to convey and symbolize much more than speech among a region's range of development actors. Talk is merely the beginning of an ongoing process of engagement and collaborative behaviors among the principal representative stakeholder individuals, institutions and organizations of the region and which affect its communities from locations beyond the region.

In order to get started soon after the local stakeholders have become aware, and thereby are sufficiently motivated to take action, the question arises, "what do we do next?" Based on the principal lessons that were identified in the earlier parts of this book, what might the stakeholders do? They can begin by organizing themselves into a strategic-level working group (see p. 106–7). The group should begin the *talk* phase of the ALERT Model process by setting its sights on at least five critical activities: (1) *Governance*: given the fragmented and uncoordinated government units of most of the world's city-regions, multiple- and cross-sectoral governance strategic planning behavior needs to be initiated and sustained for the foreseeable future. (2) *Change mindset*: this is the long-term process of changing how professional planners and constituent stakeholders think, see the economy of today and tomorrow, and how they change the content and approach of their planning for the global knowledge economy and network society. (3) *Human capital*: immediate action for setting investment priorities for short-term and long-term programs for high-quality education in the region and its localities is imperative. These actions should be driven by strategies that span the life-cycle of the forces that influence the development of intellectual capital – i.e. from early childhood development and its nutritional concerns through to graduate-school level research education, plus in-service and professional development for workers already in the workforce. (4) *Enterprise culture*: given the criticality and instrumental role of innovation and entrepreneurship in the knowledge economy, it is essential that explicit attention be devoted to, and investment of effort made in ensuring that creativity and continuous improvement become pervasive norms of behavior throughout the community and across the generations. For example, this priority may be reinforced by high-priority sustained initiatives in education, science, engineering and the arts; this is a complement to the goal of instilling a culture of creative enterprise throughout the region. (5) *Equity* – the spatial distribution of the assets and resources across any community is uneven. In order to ensure that the entire population of the region has access to, and opportunities for advancement and development, overt attention must be given to disparities of basic economic functions. By understanding the nature and locations of these disparities, planned policies and programs for addressing these gaps can be formulated and implemented.

Initial Talk Stage

The goal of the initial *talk* stage of the ALERT Model is to begin a continuous long-term iterative process of intentional and planned change for increasing the competitiveness of the region and its localities. The process may begin to be seen as increasingly successful when recognition, both internal and external to the region, is expressed that the community is progressing toward such ends as: higher-wages and fuller employment; improving quality of life; and more widespread and equal distribution of these benefits throughout the local society, both in economic terms and spatially. The principal *means* of the *talk* stage are operating under conditions of: (1) a pervasive new mindset based on increased awareness and commitment to engage strategically the global knowledge economy and network society; (2) growing higher-quality human capital investment and returns; (3) a sustained regional and local enterprise culture; (4) changed and more effective community governance behavior; and (5) improving equity status of the region's economically distressed areas. To maintain the opportunity to be competitive and successful in this planned-change process, the region's stakeholders need to recognize that this phase of the ALERT Model requires long-term sustained commitment.

In order to develop new forms and practices of local and regional governance befitting the relational complexities and requirements of the global knowledge economy and network society, the "talk" concept needs to involve sustained and continuous interaction and networking among representatives of the region's principal stakeholders. These interactions should involve and link the political economy and cultural functions and institutions of the for-profit, the governmental, and the nonprofit sectors that influence and can influence the region. Once these new partnerships and networks have imprinted on the people and organizations of the locality, the region and its communities can be better prepared, empowered and responsive to compete strategically and planfully in the global knowledge economy.

The Need for Being Continuously Alert

In this, the continuous planning and implementation of the *talk* phase of the ALERT Model, all the skills, practice, experience and emphases on social capital, relational planning and innovation must be applied and perfected on an ongoing and long-term basis. A region seeking to maintain its competitiveness will not likely see its strategic planning and plan implementation work completed. The region's planners, stakeholders and community members at large must remain aware and be planfully activist as a normal course of responsible civic behavior.

As the regional and local stakeholders get deeper into the process of the ALERT Model, they will learn a great deal about new planning tactics and new development content, such as biosciences and nanotechnology. If the interest and commitment by the stakeholder groups grow, the need for learning new actionable knowledge will also grow. Local book clubs are an important means for addressing this need. In the process, further empowerment and creativity comes with the increased knowledge. This a continuing partial response to the question, "What do we do next?" A fundamental and continuous need is to learn about other regions and the reflections of others who have struggled with issues similar to the local ones – e.g. writings on or from the global economy's three principal technology-economic and culture regions: North America, Eastern Asia, and Western Europe. Sample book club selections are listed below in the section on Continuous Learning (see pp. 169–70). For more detail on the critical planning practice activities of this phase of the ALERT Model, refer to Part V, see section on *talk* phase, pp. 233–4.

Beyond Talk: New Mindset, Governance, Practice, Equity, Surveys and Scenarios

Reflecting on the findings of Patsy Healey's analyses and informative insights from various European relational planning cases (see p. 67), several crucial observations may be made that can move the *talk* component of the ALERT Model beyond initial concern and preliminary talking to sustained planning, action and accomplishment. (1) To do this, mindset change is essential. Ways of practicing planning evolved from past conditions and influences on the region – i.e. path dependence. While some of these practices remain appropriate (Aurigi June 5–9, 2005), there have been momentous external changes that have impacted significantly many of the world's regions and localities – e.g. many local economies of "industrial" countries are no longer driven principally by manufacturing. Consequently, a break with traditional or legacy planning practice, in part, is essential. What is needed is to forge new relational planning approaches, especially ones that are buttressed in law, regulation or public mandate, as well as by new informal planning practice both by professional planners and their constituent stakeholders. (2) Routinize new cross-sectoral governance and plan-implementation practices that are regional and local collaborative relationships. In other words, these changes must be reflected in behavior as well as in mindset. (3) Development policies need to reflect the values of equality of opportunity and technology justice for all segments of local society. Operationalizing this equity value of reducing economic and technology disparities is essential in order to ensure that all segments of the regional society and economy might have the opportunity to benefit from the new-development planning and governance behaviors, regulations and potential that are congruent with the changed, and constantly changing economic conditions locally, regionally, nationally and globally.

Disparity in access to high-speed broadband technologies, which is an element of the digital divide, has been a particularly chronic problem for low-density rural areas and poor inner-city neighborhoods (Craig and Greenhill April 2005). Additional related equity concerns are found in the unevenness in the region's employment, educational attainment, income, home ownership, rent and access to quality housing. These and other equity issues, such as may be associated with race, ethnicity, gender, age and religion, require strategic analysis and planning in order to realize a local society of common, equal opportunity (cf. Provo 2002; Mayer and Provo, 2004: 23–27; Schön *et al.* 1999).

In the US, there is a highly variable tradition of regional planning, and therefore, regional governance, at the scale or layer that encompasses entire states down to metropolitan-scale planning and governance levels. Historically, in the 1930s, the Tennessee Valley Authority was a pioneer (Lilienthal 1944); it is still active today providing energy and economic development (Tennessee Valley Authority n.d.). In the 1960s, the Appalachian Regional Commission was created to address a vast region of economic and other disparities; it too is still active and also engaged in economic development, including digital development planning (Appalachian Regional Commission n.d.). These two cases are formal, national government-instigated, and largely historical examples of some of the best in US regional planning. Over the last generation, in the absence of a commitment by governments in the US to strong and effective regional-scale planning and governance, a number of more informal or multi-sectoral alliances and partnerships have emerged, and the best of these might serve as today's benchmarks for various layers of regional governance in the US. Some examples include: at the multistate layer, the Southern Growth Policies Board (Southern Growth Policies Board n.d.; Tornatzky *et al.* 2002); and at the substate layer, the Washington State Technology Alliance (Technology Alliance 2000; cf. Connect Northwest n.d.). Examination of the performance and impact of these kinds of informal regional governance partnerships can inspire planning and governance innovations elsewhere. In Canada, the Waterloo–Kitchener–Cambridge region is a multiple-stakeholder developer example (Canada's Technology Triangle n.d.; see also Healey 2002; October 20–24, 2004b).

From Actionable Knowledge to Willing Intentions into Realization

The primary concern of the book is to provide stimulation for fresh planning and strategic thinking, behavior and a fundamental change in the planning mindset, such that new approaches that are compatible with development for the global knowledge economy become embedded and second-nature in the routine practice of planning at the regional and local scale. To accomplish this, two tactics have proven to be particularly helpful: (1) surveys; and (2) planning scenarios.

There are several important realizations to be made here. Given the particular mix of assets and actors in every place, there is no one codified set of generic relational planning guidelines that will lead directly to the kind of successful competitiveness that is needed by each region around the globe. Therefore, as noted above, it means that each locality must devise its own approach. Even more to the point, each region, its localities and peoples are distinctive. In operational planning and plan-implementation terms, places are not exactly like any other. As a result, there may well be some space for the region to devise strategies that might confer some development advantage or e-business niche, thereby better positioning the locality to compete economically. Consequently, this means that particular engagement and tailoring of plans and actions for the region are required. For this ongoing activity to be effective, it is important to sense change and conditions. For us, this has meant that it is useful to conduct surveys that enable regional planning decisions to be fresh and responsive.

Surveys

For planning within the Michigan context, for example, we have sponsored periodic statewide household telephone-assisted scientific surveys. These projects sample approximately one thousand households each time that these State of the State Surveys (SOSS) are conducted. We have commissioned portions of these surveys for our research purposes every two or three years. For an example of such a recent survey project, see Wilson *et al.* 2004. Among several purposes being addressed in this project, we were able to assess Internet access, usage and attitudes of households by the regions of Michigan. See pp. 167–8 on Policies Change, for an illustration of some of the uses to which we put the survey findings from this project. Also, refer to Michigancool cities.com (n.d.) for a web-based survey approach intended to elicit perceptions of young adults of university-age on their motivations for staying in Michigan and of seeking employment outside the state.

Another example of a helpful type of survey use is the more frequent survey. As part of a bell-wether project, in Grand Rapids, Michigan, 20 community leaders were surveyed every five months over approximately a two-year time span. This approach was useful as a kind of early-warning system, since we needed to track local changes regularly and often. This monitoring approach can be most supportive of short-term regional planning efforts. For another survey application for early warning, refer to the planning scenario case study from England in Part IV (see pp. 162–5).

For the *resonsiveness* element of the ALERT Model, surveys are another means of staying abreast of, and sensing the impact of planning and the community's dynamics. Nowadays, it is cost-effective and operationally feasible to devise and execute Internet surveys as an aide to ensure that interventions indeed are addressing specific needs and the demands that have been expressed by the region's people, firms and institutions. Internet-assisted surveys also should be explored and considered for use in regularly assessing local community preferences and perceptions. A number of relatively inexpensive Internet survey options now are widely available. An example is SurveyMonkey.com (n.d.).

We also have used traditional mailed surveys combined with e-surveys to assess the degree to which professional urban and regional planning practitioners have changed their mindsets and practice behaviors within the context of enabling their regions and localities to prosper in the new environment of the global knowledge economy and network society (Breuckman April 2003; Singh July 2003).

Planning Scenarios

Engaging in the planning of scenarios for the future of the region and its localities is highly recommended (Ogilvy 2002; Fahey and Randall 1998). This is an effective way of following up on the awareness that has been attained from the opening stage of the ALERT Model process. Constructing realistic planning scenarios is a useful means of focusing systematic and collective attention on various planned futures for the area (Ringland 2002). Refer to Part V, section on Relational Planning Concepts, A–Z; see Planning Scenario, pp. 212–13. Compare this to Scenario in the same section (p. 216). Planning scenarios facilitate discussion, experimentation and testing of various ideas and approaches for future development. For ideas on future activities, see University of Arizona n.d.

A number of benefits can result from making scenario planning a keystone of the *talk* stage of the ALERT Model. First, and most importantly, scenario planning can produce substantive ideas, economic-content options, and a range of innovative and feasible planning solutions for use in the actual development of the city-region. Planning scenario activities are ways of learning about and practicing intelligent-development planning. Second, in order to initiate, practice and construct new governance behavior, and in the process to invest in social capital formation among key actors of the city-region, it is useful to conduct scenario planning activities among the community's representative stakeholders. Third, in the process of conducting scenario planning, new learning will occur that may serve to empower regional stakeholders because they will have advanced their understanding of both the content objectives and the process of practicing strategic relational planning. This represents the initial development of a learning community for the region. Over time, the more progress and success that these efforts produce, the more likely the overall regional strategic planning effort will become widely known and accepted throughout the area's population and its institutions. How might one go about doing scenario planning?

We have tested the applicability of the ALERT Model and some of the relational planning concepts underpinning the Model. To do this, we have constructed and can share commentary on eight planning-scenario cases. The first examples are from the US. There are four cases from Eastern Asia;

two of these span the national economies of Southeast Asia and two others are applied to South Korea. Finally, there are two cases of planning scenarios from Western Europe. This diversity of cases depicts a variety of starting points and ending points for the participants who engaged in the planning scenario exercises.

Cases of Planning Scenarios

The goal of this section is to illustrate the usefulness and the different kinds of planning scenarios that can be used to test the feasibility of various relational planning ideas inherent in the ALERT Model. The planning scenarios vary by purpose, scale, location and technology-economic context, among other variables. Scenarios 1 and 2 both stem from our relational planning practice work with the East Central Michigan planning region. This region is typical in that it is not advantaged by being in close proximity to the metropolitan areas of Detroit and Grand Rapids. Further, the general empirical context for these planning scenarios is helpful also in that this case region is representative of many economically developed regions around the world, the local economies of which were driven primarily by manufacturing industries.

A range of empirical settings in Eastern Asia has been used to test the generic applicability of the ALERT Model. Each case has demonstrated different dimensions of the relational planning task. Scenario 3 used the product life-cycle model to show how Singapore-based firms might plan to offshore some of their production functions throughout various locations in the Southeast Asian region and Asia in general. Scenario 4 demonstrated how the countries of the Southeast Asian region might enhance economic development by means of good governance and e-governance. Scenario 5 is a case of regional planning for the Seoul-region suburban Gyeonggi province of the Republic of Korea. Scenario 6 is an example of scenario planning that can be useful in generating alternative planning strategies; in this case, the planning scenario is an alternative to the South Korean president's plan to relocate the capital of the Republic of Korea. Two European planning-scenario cases include discussion of a regional and local-level planning-scenario case study from England – i.e. Scenario 7 – and a continental scale case study of likely futures scenarios for the enlarged European Union – i.e. Scenario 8. These two examples are included here because they enable the reader to get inside and obtain an operational understanding of several methods and techniques for producing different types of useful scenarios for planning. The eight scenarios were derived from the three major technology-economic regions of the globe that include several of the world's most innovative technology-based economies for the planning, widespread distribution and access of digital infrastructure throughout their respective national areas.

Scenario 1: Planning Scenario for Training Regional and Local Planners in East Central Michigan

In order to pilot this part of the ALERT Model, a Planning Support Team engaged in planning-scenario development activities using the case of Michigan's East Central Planning Region. To place it generally, this region includes the tri-cities of Saginaw, Bay City and Midland, Michigan (see Figure 19, p. 120). The global headquarters of The Dow Chemical Company is located in Midland. The six economic development theories that were introduced in Part II (see pp. 52–3) were used by the Planning Support Team to advance their actionable knowledge. In conducting this exercise, the notion was that once the Planning Support Team members became comfortable and facile with executing the planning-scenario development process, they would be better positioned to work with the area's other regional and local planners, along with the community's stakeholder representatives, to empower them to engage in the processes of strategic planning innovation and creativity as they bring together the economic development theories with the actual empirical realities of the region and its localities. Given the particularities of places, it is essential that the area's representatives devise and invent their own approaches to their special futures as they envision them – short-term, medium-term and long-term. Once the Planning Support Team is prepared, it may engage and mobilize the region's planners and stakeholders more effectively. Further, the initial groundwork will have been laid for diffusing this planning-scenario learning by means of a "training the trainers" approach, whereby the Planning Support Team can train the Regional Strategic Planning Steering Group of stakeholders, and in turn, the Steering Group can train its Working Subgroups, and so on, as the needs arise.

Charrette Workshop Process

A two-stage process was recommended as a result of the training process: (1) the Planning Support Team conducted its own experimental charrette workshop, and (2) the team then was positioned to lead the Regional Strategic Planning Steering Group of stakeholders and its Working Subgroups through tailored, but similar charrette workshops. Thus, an initial straightforward pathway to the continuation of this learning and planning practice process is to assign a strategic planning group of stakeholders, each of the same six economic development theories to the working subgroups of the overall group. In turn, each subgroup would work from and with its assigned development theory with facilitation from members of the region's Planning Support Team. From these experimental learning activities of the group and subgroups they need to practice strategic relational planning. Their ongoing engagement and resulting experience will enable them to perfect and connect empirical facts and strategic planning alternatives to the required layers and functions of the economy – local to regional to national and to the global economy, using the e-business spectrum as a template to check that all of the region's principal economic functions are considered. For example, a basic action context for these exercises is to ensure that both the digital development infrastructure requirements and the intelligent development content potentials noted in Part IV are in place and underway simultaneously.

These various strategic planning activities may be organized into charrette workshops among the respective working groups. In the case of the Michigan East Central Planning Region, the local regional planners had taken the lead in initiating a strategic planning process. In order to train itself in relational planning practice and in an effort to provide those local and regional planners and their stakeholders with a diversity of strategic options, the Planning Support Team met in workshop format and developed planning scenarios that might be considered in the initial and the follow-on development of various planning options for the region's future.

Early in the charrette workshop process, the pre-workshop homework assignments were discussed briefly. The discussion covered each of seven assigned theories or models:

1 Growth poles and growth centers (including spillovers and spread).
2 Product-cycle model.
3 Industry life-cycle.
4 Flexible production and flexible specialization.
5 Learning regions and innovative milieux.
6 Competitive advantage.
7 Enterprise segmentation and unequal power relations.

As assigned, each participant came to the workshop with prepared notes and briefing material on their respective assigned concept. This facilitates the efficiency of completing the charrette activities quickly. Such preparation also helps the participants to focus and concentrate on their respective contribution to the overall task of generating conceptually informed alternative planning scenarios. Any needed clarifications and revisions of these materials were requested to be distributed to the team immediately after the workshop. A few references were requested to be added to the briefing materials, so that readers might do additional follow-up. The Planning Support Team version of the charrette requested that the staff participants put their notes and briefing material in Word-document form to assist in electronic sharing of the information before and after the charrette workshop.

During the workshop, the substantive result was a diverse list of empirically-based and theory-informed actions that initially were simple content and action items. The generation of these items served both to produce content priorities and possible future planning directions, but also importantly, the exercise enabled the participants to engage in and begin the process of planning to determine the future of the region. Breaking out of a cycle of passively receiving plans from on high is critical to the local empowerment and learning process. Simply, this is the beginning of the local stakeholders and their planning-support staff and volunteers being able to answer their own question of "Now what do we do?".

Results and Priorities

In the case of the initial Planning Support Team's pilot charrette, 43 issue responses were generated. The team was able to take the time to prioritize, vote and produce a ranking of these issues. Two issues received the team's highest priorities: the number one issue, by far, was a merger of two closely-related organizational and governance concerns; both were intended to catalyze the region's strategic planning process. The team wants to see the formation of a neutral organization to plan and coordinate the region's strategies for future development. The notion was to avoid the legacies of the existing organizations, which have combined to bring the region to its present state. The new governance entity should span sectors and organizations; it should be a non-profit organization that can raise funds and receive grants, donations and gifts. The second dimension to this merged issue was the need for the region to form a public–private business consortium or alliance to promote innovation and to generate science and technology-based investment throughout the region. The number two ranked issue was to ensure awareness throughout the region of its standing in today's economy and to ensure further that this information leads to the mobilization of stakeholders to strategic planning and planned action. The number three issue called for the area to form a learning region by means of strategic planning. Twenty other issues were listed, but their respective weights in the voting procedure did not reveal high priority for any one of these other responses. However, with the exception of multiple-issue interests related to the biosciences, the range and diverse mix of the many other issues did reveal a listing of content and procedural issues that will be useful in designing and operating future stages of the region's intelligent-development planning. For example, a selection of a few content issues for future development included: the priority to assess the region's biosciences strengths and depth; in turn, there was the related priority to strengthen and frame an educational response to the biosciences and capitalize on bioscience strengths for development of the industry and its products; reform education in the region; and capitalize on the agricultural resources of the region's localities. The other ranked, largely procedural and spatial organizational issues included the need to:

- Consider the establishment of a future-technology management institute – i.e. an in-plant approach with a mix of industry and higher education with best local practitioners as mentors and instructors.
- Assess the region's current identity and develop a long-term future spatial identity that is more reflective of the locations of intended planned economic functions. These may be more polycentric and thereby more reflective of needed and planned future linkages and realities.
- Structure the region's future spatial organization over the next 25 years to be polycentric. This will consist of: (a) individual monocentric cities and towns; (b) the North-South corridor city; and (c) the network city.
- Establish a technology-innovation network organization, the purpose of which is to expand, deepen and spin-off innovations based on the region's biochemistry potential as represented by the current chemistry industry and the area's agricultural base.
- Raise the international awareness for leaders in the region and compare the region to the more than seventy learning regions of the European Union.
- Identify and appoint a champion to lead selected development programs by sector – for example, the development of biosciences specialties. Champions for several sectors may be needed, especially in the early stages of a new sector's development.
- Prioritize, as needed, coordinated private and public investment in selected industries based on product life-cycle status.
- Assess potential for the establishment of flexible production and specialization districts for the region; reduce the transaction costs for these firms.
- Devise an effective business-retention strategy to maintain comparative and competitive advantage, both long-term and short-term.

Continuation of Planning Scenario Development for the City-region

These issues are exemplary of the kinds of thoughtful would-be action items from an initially informed professionally supported stakeholder-based strategic planning process event. Except for some follow-on research, the Planning Support Team ended its involvement with the East Central Michigan region at this stage. The next concern would be to move these awareness-building activities to acting on the next phases of planning. These actions would involve moving the process toward rigorous and data-based plan-making and plan-implementation. The remaining stages of the ALERT Model would be able to serve as the framework for such follow-on action. See the Isle of Wight, England, scenario case described on pp. 162–5 for an example of a local-scale planning process that was fully implemented. Had this East Central Michigan planning scenario been continued, one of the next set of activities would have been to extend the process, as updated and revised, to communities throughout the planning region. The Planning Support Team then would have considered and possibly designed an electronic charrette process application (Paul 1997). This would have enabled a more widespread and less expensive means of generating creative development-planning solutions and broader ownership of the planning process and its outcomes.

Scenario 2: A Research Planning Scenario for the East Central Michigan City-region

As a result of the exposure of the Planning Support Team to the strategic development needs of the East Central Michigan region, one of the team members, Karan Singh, who was a graduate planning student, designed a planning research project to explore the region's potential as a location for the further development of its biosciences sector (Singh 2004). The project's research question was: "In Michigan, are the biosciences likely to benefit regions external to the large metropolitan cities and the Life Sciences Corridor, which have an existing built up mass of bioscience firms?"

The operational definition of the biosciences used by Singh was more encompassing than the typically narrow definition of biotechnology. Singh's "biosciences" included the life-sciences industry; it "involves biological R&D, the production of products required for innovative scientific and medical procedures, and the practice of advanced medical treatments" (Anderson Economic Group 2004). Refer to Part V, section on Relational Planning Concepts, A–Z; see Biosciences and Biotechnology industry, pp. 190–2. Singh's research employed the method of developing planning scenario*s* for the East Central Michigan region. He developed the planning scenarios by applying the Program Planning Model (PPM) to the issue of the feasibility of that region hosting a successful strategic planning strategy focused on developing an enhanced biosciences sector (cf. Corey 1988; Van de Ven and Koenig, Jr Spring/ Summer 1976). Refer to Part V, see two sections, one entitled an "Outline of a Planning Scenario Approach . . . by Karan Singh" for an outline of the Program Planning Model's application (pp. 235–6); also see the generic description of the model, entitled, "A Time-Relational Method: The Program Planning Model, (pp. 237–9)." Within the Program Planning Model context, life-cycle theory informed the development of the planned strategy. Three planning scenarios were generated by Singh from the analyses and interpretations for the development of the biosciences in Midland and Isabella counties in the: (1) near-term – target years 2004–7; (2) medium-term – target years 2008–15; and (3) long-term – target years 2016–25. Even though the region is located external to the Michigan Life Sciences Corridor and the state's two largest metropolitan regions of Detroit and Grand Rapids, the planning-scenario research enabled Singh to conclude that such a biosciences development strategy was feasible for the stakeholders of the East Central Michigan region to consider in their future strategic planning.

This particular application of a planning scenario methodology was concluded to be:

> a thoughtful and measured approach towards planning for a biosciences economy within the two counties of Midland and Isabella [of the planning region]. The scenarios are rooted in the implicit realities of the counties existing bioscience strengths, while trying to build upon those economic strengths and plan for future economic development potentialities.
>
> (Singh 2004: 143)

In the planned pilot stage of the planning scenario, Singh included a design for measuring the program's success. The program design also included an iterative evaluation process.

This case in the use of a planning scenario approach demonstrated yet another way by which several of the concepts and theories introduced in Part II can be used in systematically testing the feasibility of relational planning alternatives. Note, the two e-business spectrums of policies impacting the local region (see Figure 24, p. 127) and the blank template of the e-business spectrum (see Figure 25, p. 128) may be useful in this planning scenario case, as means to stimulate new policies, programs, priorities and phasing for the various planned scenarios.

Scenario 3: Applying Spatial Economic Development Theory to Southeast Asia

The purpose of developing this planning scenario was to test selected elements of the ALERT Model. Drawing from the Singapore context and strategic planning perspective, one may demonstrate a simple example of a development theory application at the macro-scale layer of a technology-economic region of multiple countries such as Southeast Asia.

Tiny Singapore's scarcest resource is lack of space for development. The life-cycle concept suggests that products and industries often demonstrate staged development over time, such that one might identify: a formative stage; an intermediate-growth stage; a mature stage; and a stage of decline that also may stimulate a new formative stage of extended or related new development (cf. Audretsch and Feldman 1996). The principles of the industry life-cycle theory and spatial relational planning, for example, can be useful in planning for those technology-based production activities that do not require proximity and concentration. Such economic functions therefore may be located in Singapore, but outside of the city-state's planned prime-location technology corridors (Corey 2000). Alternatively, these technology-based production functions may be located outside Singapore all together. In the case of the latter, development can be "intelligent" by building explicit twinning relationships such as between research and development (R&D) institutions and startup companies in Singapore and functionally related production companies in, for example, Johor state or the Multimedia Super Corridor in Malaysia, or in Indonesia, Vietnam, India or China. This would reserve precious technology-corridor space in Singapore primarily for the front-end innovative intellectual-property oriented value-added production functions that must be concentrated with R&D affinity activities that require face-to-face proximity for the exchange of tacit knowledge. In contrast, some of those production and consumption functions that have evolved to the stage when they rely principally on codified information activities need not continue to occupy the limited space of the technology corridors (Corey 2004a). This may mean that it is more space-effective and cost-savings efficient for a firm and a host country to construct scenario plans to relocate certain of its products and its industries to reflect the life-cycle stage of their development. For a spatial representation of the various life-cycle stages and their intended locations, refer to Figure 10.9, "Life-cycle Relational Spatial Planning: A Southeast Asian Scenario," in Corey and Wilson, 2005b: 335. This is an example of a planned scenario of "intelligent development" because it recognizes that a one-size-fits-all approach to locational analysis and spatial planning is too simplistic for today's complex and multidimensional global knowledge economy and network society. Singapore's policies of localization, regionalization and globalization strategies (Corey 2004a) are accommodated by the ALERT Model's multiple layers of relational potentials and power of agency relationships that occur at different scales – both internal to Singapore and external to the city-state as its policies seek to orient business initiatives toward the international marketplace (cf. also to the space and time element of Graham and Healey 1999). Refer to Figure 22 on p. 124 for an illustration of Southeast Asian Strategies Layers.

In the field of economic development and spatial planning, there are a number of other related theories and concepts (Plummer and Taylor 2001a) that have undergone empirical testing and the establishment of the validity of these models and measures from application to regions in Australia (Plummer and Taylor 2001b). As introduced in Part II, Plummer and Taylor have examined six largely spatial theories of local and regional economic development; these include:

- Growth poles and growth centers (spillovers and spread).
- Product-cycle model (industry life-cycle).
- Flexible production and flexible specialization.
- Learning regions and innovative milieux.
- Competitive advantage.
- Enterprise segmentation and unequal power relations.

Each of these models is defined in Part V; see section entitled Relational Planning Concepts, A–Z, pp. 189–220.

As was suggested by the simple example of the application of the industry life-cycle theory noted above in this planning scenario, the body of theoretical work can be helpful to corporate and government strategic planners in taking action within the context of the ALERT Model and in operationalizing Graham and Healey's guiding principles on the multiple meanings of space and time. This kind of concept utilization also presents potential for constructing other ICT-facilitated or digital development and intelligent development planning scenarios. A case of addressing the development of good governance in Southeast Asia as part of a planning scenario application is illustrated in the section that follows.

Scenario 4: Toward Good e-Governance in Southeast Asia

Planning scenarios lend themselves to being able to position better regional and local plan-making to facilitate monitoring and assessing anticipated planned results. Looking to the future, effective relational planning therefore requires laying the groundwork for evaluating future policy outcomes. By integrating this purpose with the relevant theories, the principles of good governance, intelligent-development policies and effective planning, the necessary seeds may be sown to go beyond digital development and on to successful planned intelligent development implementation. In Southeast Asia, improving governance has been an issue, especially within the context of the inclusion of civil society in doing effective and successful development planning. Since this region has several of Asia's more progressive economies regarding their embracing of ICT-enabled planned development – e.g. Singapore and Malaysia – there is value in projecting likely scenarios into the future as digital development and ICT infrastructure throughout Southeast Asia become more accessible.

This scenario construction demonstrates some aspects of the time element of Graham and Healey's relational planning practice scheme. We have monitored ICT- and science and technologies-based policies and programs in the Southeast Asian region for nearly 20 years. As a result, we have been able to form some longitudinal assessments (Corey 1998; Corey and Wilson 2005b). From these snapshots from the near past, one may attempt to project into the future some general development expectations.

Setting the Context

Empirically, the country-economies of the region have organized themselves into four technology-economic worlds. (1) Singapore occupies the region's first world. Its people had the foresight, leadership, financial resources, planning innovation and culture, priority and earliest policies implementation to distinguish its digital development and intelligent development regionally and globally. (2) Malaysia and Brunei inhabit the second world of the region; Brunei is placed here because its relatively wealthy economy from petroleum revenues has enabled it to be a performer on some dimensions of digital access, but not in terms of intelligent development content and applications. Malaysia emulated much of the progressive technology-stimulated development of near-neighbor Singapore, and somewhat later than Singapore, moved rapidly on engaging the forces of globalization for development, and moved on implementing bold long-range visions of intelligent development for some areas of the country, especially as exemplified by the Multimedia Super Corridor, which includes a new administrative capital, mega-project infrastructure and the championing of these intelligent developments by leadership from the top of government (Bunnell, 2004; Bunnell *et al.* 2002). (3) The region's third world of technology-economic performance is occupied by Thailand, the Philippines, Indonesia and to some extent, Vietnam. These countries were badly hit by the Asian financial crisis of 1997–8, thereby greatly slowing their plan-implementation for technology-based development. (4) The fourth and bottom world of Southeast Asia's technology-economic performers include Cambodia, Laos and Myanmar. These countries effectively remain unconnected to the global knowledge economy and network society. This layering of technology-facilitated development results for the region has remained stable and consistent over the decade of assessments that we have conducted. If this structure of digital and intelligent development "haves and have-not" is to be unfrozen, then macro-scaled regional strategic policies and program planning, and active plan implementation are imperative.

Action Framework and Elements for an e-Governance Planning Scenario

Within the *talk* and *beyond talk* phases of the ALERT Model, five interdependent elements have been derived to be used in informing the construction of a planning scenario for the region's technology-economic future. Three of the elements are ends, or should serve as goals of the planned change process; two of the elements are means, or should serve as ways to achieve the goals. The three goals are: (1) equity; (2) quality human capital development; and (3) enterprise culture of innovation and entrepreneurship. The two means or methods to be used to work toward achieving the goals are (4) mindset change and (5) governance.

Equity

Inspired in this context by the layering of the digital-development disparities that have been noted above in the layering of the four digital worlds for Southeast Asia, these digital divides should be addressed so as to pursue equity by reducing the disparities. Reinforcing this theme, economist Catherine Mann has written:

> That it is valid and useful to consider developing nations as distinct from developed countries in terms of their economic status and the ICT con-

> ditions. Differences in income per capita define developing countries and distinguish them from high-income industrial countries, and as a general statement, the digital divide of ICTs exists just as does the income divide. Capturing the diversity and unity among the developing countries is also an essential component for understanding and improving the contributions ICT can make to international development; given the right policy and institutional environment.
>
> (Mann 2004: 76)

This close relationship between income and level of technology reinforces the critical role that the economy plays in regional and local development. Consequently, the next two elements are objectives, that if successfully pursued ultimately should contribute to the narrowing of the economic and technology divides.

Human Capital and Enterprise Culture

As noted in Part II, see section on "Space: Theory Integration, Benchmarking and Measurement," pp. 54–7, Corey and Wilson have used Plummer and Taylor's Australian regional analyses to derive the composite model – i.e. the ALERT Model (Corey and Wilson 2005b). It recognizes the primacy of human capital development and enterprise culture in producing successful local economic development within the context of the global knowledge economy (Plummer and Taylor 2001b). There are other related concepts – e.g. path dependence – that can be used also to stimulate local awareness and planning scenario construction for future digital and intelligent development. Refer to Part V, see section on Relational Planning Concepts, A–Z; see Human Capital and Path Dependence, p. 203 and pp. 211–12.

Governance and Mindset Change

For development planning aspirations and intentions to be realized, it is at the regional and local level or layer that attention needs to be directed. This is the primary scale at which planned change and action must occur. Planning scenarios constructed at the level of the locality can serve as a complement to technology-supported policies and programs from national and international levels. For example, the international regional organization, the Association of Southeast Asian Nations (ASEAN) began an "e-ASEAN" initiative as long ago as 1999. For the Southeast Asian region, this initiative: "Aims to develop a broad-based and comprehensive action plan including physical, legal, logistical, social and economic infrastructure needed to promote an ASEAN e-space, as part of an ASEAN positioning and branding strategy" (Association of Southeast Asian Nations n.d.). The e-ASEAN initiative has included an explicit concern for narrowing the digital divide in Southeast Asia. To date, the focus of the initiative has been on e-commerce and has involved pilot projects. The effectiveness and impact of these top-down activities has yet to demonstrate noticeable impact (cf. Corey 1998; and Corey and Wilson 2005b). By means of local, bottom-up planning scenario application, in concert with ASEAN level and national level coordination, it may be possible to more effectively attain these aims of the e-ASEAN initiative. This kind of *alignment* of investment, intention and continuous commitment over the long-term can be effective in producing the planned changes stated in the e-ASEAN action plan cited above. Refer to Part V, section on Relational Planning Concepts, A–Z, see Alignment, p. 190.

The local region and its communities comprise the scale that directly impacts the lives of people and the performance of firms and other organizations. Given that governance is about enhancing the working relationships among the development interests of civil society, private-sector firms and institutions, and governments, there is value in planning within the e-ASEAN context for regional stakeholders to develop alternative development planning scenarios for select localities of the region. For operating definitions of "governance" and "good governance," refer to Part V, section on Relational Planning Concepts, A–Z, pp. 201–2.

This development planning will be influenced and limited by the state of ICT infrastructure and by where the locality is positioned in this digital development context. Such planning will be assisted by systematically graphing and mapping the distributions and levels of digital development layers and spatial organization throughout the Southeast Asian region. The point to begin realizing the e-ASEAN vision is by means of local pilot efforts. Executed with intensity and sustained dedication to plan-implementation, subnational and local efforts of digital development and intelligent development planning scenarios have the potential for catalyzing change and ultimately leading to less development unevenness across the region. As local digital development infrastructure, access and use might increase, the capacity for intelligent development will have been seeded in, both for the present and into the future, for realizing sustained long-range technology-enabled growth and enrichment.

The practice of new-governance behavior is interdependent with changing existing mindsets among the development actors. The more local stakeholders experience and practice new governance, the more likely planning mindsets will be changed and vice versa.

Government and governance are not the same. For Asia, this distinction may be particularly important. When compared to English-speaking countries, Michael Minges of the International Telecommunication Union noted that Asian governments took the lead in promoting Internet access, while anglophone countries relied more on the private sector (International Telecommunication Union November 19, 2003: 1; Williams November 20, 2003). There may be a tendency to wait for, and expect government leadership to address economic and digital divides, as well as digital and intelligent development. This reliance has influenced the uneven digital development in the region (Corey and Wilson 2005b). Similarly, the performance to date of government-led e-ASEAN has been modest and has not yet demonstrated the impact required to close the economic and technology disparities of Southeast Asia.

New subregional and local governance goes beyond the concept of the state or of government per se (cf. Madanipour, Hull and Healey 2001). If some significant digital and intelligent development inequities are to be overcome, it is incumbent on the localities of Southeast Asia also to overcome their tendencies to wait for government to lead on these issues. Just as some of the digital and intelligent development leaders of the region – e.g. Singapore and Malaysia, took the initiative, localities and city-regions in the other countries of Southeast Asia should consider taking similar initiatives. They might well begin by developing relevant and tailored planning scenarios similar to those regional and local cases that are illustrated in this part of the book. The parallel organizational and planned-change technologies and approaches elaborated in these selected case studies might be especially useful – e.g. the European applications, especially the Isle of Wight, England, case; the Michigan city-region processes; and Gyeonggi province, South Korea. From these local planning practice examples, planners and stakeholders across Southeast Asia may be positioned to organize their planned development by adopting the spirit of relationality of the ALERT Model processes to invent their own particular engagement with the potential and dynamics of the global knowledge economy and network society.

The pincer of integrating the current top-down approaches with the just-suggested bottom-up approaches can provide the stimulus and mobilization required to produce greater development impact. In the process, the disparities and divides within and between the economies of the region may begin to be bridged more effectively.

Policies Planning and Policies Evaluation

As local planning scenario construction might proceed throughout the less-developed places of the region, it would be important to incorporate best practices and benchmarking research on governance in the developing world, see the Development Gateway (n.d.) website. Indicators of governance and institutional quality may be found on the website of the World Bank Group (2001). Also, refer to the United Nations Development Program (UNDP) which has provided a list of behaviors and characteristics that can be used as a checklist to help frame and use democratic governance indicators for constructing policies, program objectives and producing results for good governance scenario planning (United Nations Development Programme 2004).

In the development of local planning-scenarios, the operational details of designing the means for monitoring and evaluating the implementation of the policies that are generated from the local planning-scenario activities need to be explicit and crafted carefully. It is critical that the data and measures used reflect and relate directly to the planned policies and the execution of those policies (see the section on Policies Change pp. 167–8), and refer to Corey and Wilson 2005b).

Finally, a practical and effective way to activate subnational and local initiatives across the region is to stimulate widespread use of digital infrastructure, especially for enhancing everyday life and business. This means not only greatly expanding e-commerce and e-government opportunities, but also speeding up and expanding the rollout of ICT infrastructure so that larger proportions of the region's population of residents, firms and institutions have access to, and support for improving the use of, and content development by means of these technologies. In this context, it is important to explore and consider the latest consumer ICTs such as wireless broadband and mobile personal technologies. One of the strategic uses of these technologies that should be explored includes ICT applications to develop planning scenarios for good governance. Such e-governance would be an important way to practice and support the extension of both digital development and intelligent development to unserved and underserved regions throughout Southeast Asia. Working on the joint economic and technology disparities both up from the grass-roots local level and down from the regional and national levels – i.e. by ASEAN and by each individual country respectively – may be more effective than continuing to rely primarily on the current top-down national government-driven strategy approach. Such a proposition may be initiated by conducting a planner-led planning scenarios project for representative stakeholders at the region-wide level, such as the macro-scaled "Future of European Regions" case described below as Scenario 8 (see pp. 165–6), and then by following up that exercise at local and subnational regional levels by developing tailored micro-scaled scenario planning for the particular places whose stakeholders are sufficiently aware, motivated and responsive to pursue the ALERT Model for their own local future development planning (cf. Ravenhill 2001; and I-Ways 2005).

In sum, by using several layers or scales of planning scenarios, regional and local planners may test the short-term and long-term feasibility of developing new governance and good governance relationships locally, and in the process begin to change strategic planning mindsets. In turn, these new relationalities may be able to inspire substantive scenario planning for technology and economic equity, and for high-priority investments to be sustained for the continuous development of human capital and the creation of a regional and local enterprise culture. These five interdependent relational planning elements thereby may assist practicing planners to activate the TALK phase of the ALERT Model. In Southeast Asia, if these localized relational planning elements are couched and pursued within the macro governance context of ASEAN and the government context of their respective country, and in local area cross-sector collaboration, sufficient mobilization may occur to enable a new good e-governance approach that have e-ASEAN goals actually realized more widely throughout the region.

Scenario 5: Suburban Regional Planning – The Case of Gyeonggi Province, Republic of South Korea

Gyeonggi province in South Korea is a suburban region nearly surrounding the city of Seoul, the capital of the Republic of Korea. Similar to many other suburban places around the world, Gyeonggi province is an area in need of community identity. Gyeonggi province was used as a case study for exploring the construction of a theoretical framework that is appropriate to informing the practice of suburban regional planning within the context of the new dynamics of the global knowledge economy and network society. Further, it was intended that the planners of the Gyeonggi region should be able to consider tailoring and using the framework and its underlying principles as they engage in planned place-making for this dynamic and growing region. An additional objective of the project was to go beyond conventional definitions of "environment" and "ecology," and to integrate the Korean concept of "Salim" or life-culture into the application of the framework. Gill-Chin Lim has defined life-culture as one "In which [the] life of every human being is cherished, [a] culture in which life sustaining human activities are carried out with a sense of living in harmony with others and nature" (Lim December 19, 2003: 3). The life-culture concept is fully congruent with relational thinking. For example, Henry Wai-chung Yeung has discussed the "explanatory power in socio-spatial relations among such actors as individuals, firms, institutions, and other nonhuman actants" (Yeung March 13, 2002: 2). From the perspective of contributing to the development of the ALERT Model, the regional planning of Gyeonggi province was an opportunity to incorporate the environmental and ecological cultural functions into the ALERT Model building.

The method used to communicate the model's development was to introduce its principal elements, that when integrated were intended to enable Gyeonggi province to formulate strategic plans for its intelligent development within the context of the global knowledge economy and network society. Inspired by Graham and Healey's four-element relational planning-practice scheme, these elements were developed as generic issues for operationalization as follows:

Relations and processes. The e-business spectrum functions as a relational spatial organizing framework.

Space and time. Space: Plummer and Taylor's review of the six spatial economic development theories (Plummer and Taylor 2001a) were the basis to derive two instrumental processes that explain success in regional economic development; these were: (1) local human resources and human capital development, and (2) the development of a local enterprise culture. Time: the relational Program Planning Model (Van de Ven and Koenig, Jr Spring–Summer 1976), and the importance of the moment in practice.

Multiple layers. Using the case of the state of Michigan, various spatial layers of power relations were used to demonstrate the way that Gyeonggi province stakeholders might identify and develop their own particular layers of power relations.

Mediation. Between the layers of the power geometries of place – i.e. guided by select economic development criteria – four scenario options for responsiveness were identified and two power mediations were framed; these would-be actions were motivated by the values imperative of "development" that included an explicit equity value that should be sought by means of planning.

With these generic conceptual elements and derivations having been laid out, it was important next to connect to the issues of Gyeonggi province, especially from the perspective of planning influenced by the Korean life-culture concept. Consequently, the region's planners and stakeholders were encouraged to invent a Korean and Gyeonggi approach to intelligent-development planning informed by these previously-introduced concepts and framework. Within this context, several questions were posed for Gyeonggi planners and leaders to answer. What are the means for developing and maintaining the identity of the province? How can Gyeonggi province's planners secure the trust of the people, strengthen professional expertise and have accountability? How can science and technology, and information and communications technologies be used to reinforce a sense of community and improve governance?

For the purposes of illustrating a benchmark case, Singapore was suggested because of the city-state's recent experience in inventing innovative approaches for identity planning that feeds into the formation of the country's national Concept Plan. Singapore was offered for benchmarking purposes also because of its innovations in creating a vibrant arts, media and culture sector for the local economy and within the goal of developing a more attractive physical and stimulating cultural environment. Best practices examples from Western Europe were referenced. Five lessons were derived from that comparative body of early attempts at "strengthening regional identity and developing new forms of regional collaboration" (Corey December 19, 2003: 85).

The need was identified to stimulate the new leadership and visioning needed to transform the life-culture of Gyeonggi province into a coherent region that would be attractive and competitive in the context of the complex and continuously changing environment of the global knowledge economy and network society. The concept of development noted above, that includes objectives for reducing economic disparities and improving equity, was proposed for the core vision of the province's future vision. The relationships between this development concept and the life-culture concept were observed to be compatible and mutually reinforcing. The future vision should address the stated concerns for enhancing the Gyeonggi region's identity and the sense of community and trust among the community's stakeholders, along with the creative application of ICTs and science and technologies to advance the development content and functional priorities of the province.

There is the need to stimulate the regional strategic planning of Gyeonggi planners and the province's stakeholders to connect the above elements of visioning with the conceptual framework that was constructed from generic theory and development experience elsewhere. These operating principles, lessons and planning-practice behaviors were offered to be considered for the "Gyeonggi-ization" of the future regional strategic planning of the province. This represented an explicit list of planning tactics to be considered that incorporated the cultural and relational Asianization values cited in Part II by Gill-Chin Lim and Henry H.W.C. Yeung. The outline of such tactics included: the need for the full localization and tailoring of the region's unique life-culture circumstances now and into the future; highest strategic priority to be considered for the province should be given to the development of human and intellectual capital; building the capacity of the institutions of the region's nonprofit and nongovernmental sector with priority devoted to the arts, culture, the media, communication and the sustainability of the natural environment; and invest in the realization of an innovative enterprise culture on behalf of the province. A means of addressing this is to focus on adopting and promoting a culture of plan-implementation – e.g. this would involve benchmarking similar to Singapore, by ensuring that Gyeonggi province's digital and ICT infrastructure, as well as its traditional physical infrastructure, are integrated strategically and kept at contemporary cutting-edge levels. This should facilitate the capacity and reinforce the will of the provincial policies and programs simultaneously to emphasize planned digital development and the content of intelligent development strategies that are coordinated and implemented by a range of actors and stakeholders across the region.

Finally, in order to be effective in employing and achieving these suggestions, Gyeonggi provincial stakeholder-planners will have had to address the incongruity of past and existing relevant laws and regulations. Unfreezing and breaking out of the path dependency cycle of legacy planning will be contingent on results from the successful practice of relational planning. This will mean establishing new strategic planning behavior and practice that are mandated by new laws and regulations – i.e. ones that are facilitative of the new and constantly changing dynamics of the global knowledge economy and network society. Changing old mindsets to practice new relational planning also is essential. For example, perceiving the region of Gyeonggi province as a single unitary spatial organization or "a geography" is part of the old mindset. What will be needed is to strategize and plan development scenarios that reflect the many current and future, multilayered geographies of the relevant economic and life-culture functions of the province. In addition to its various levels and scales of economic functions and their diverse spatial organizations, they have their own locational elements of nodes, flows, networks and boundaries, most of which are not coterminous. These spatial relationalities are reflective of today's geographic complexities. Planning scenarios for the future also must incorporate analogous complexities for the various combinations of time horizons and functional intentions. Given the configurations of the Gyeonggi region's existing government space – e.g. a nearly ring-shaped political geography and being a polynuclear suburban development to the huge city-region of Seoul, there is a body of empirical studies and interpretations that may be partially stimulative when doing relational planning for Gyeonggi's intended future geographies (see Hall *et al.* July 4, 2003; Urban Institute Ireland n.d.; Thrush n.d.; Lang 2003; Garreau 1991). Planned future spatial organizations also would have to be guided by the known tendencies of concentration and clustering, and by those economic functions that behave in deconcentrated and dispersed spatial organizational patterns (Kellerman 2002).

In the end, for this project, the life-culture aspirations of the people and institutions of Gyeonggi province need to be engaged, and from that engagement, planned strategic actions need to be formulated. In the process, widespread ownership in these new future visions may be realized. Explicitly integrating the development from the tradition pathway (Goulet 1983: 19–20) – i.e. Korean life-culture values into the new relationalities of the global knowledge economy and network society – can operate to stimulate the creativity and innovation that will be required for the realization of the distinctive local ecological, environmental and quality-of-life influenced aspirations of happiness, healthiness and harmony. The province is not alone in its desires to realize these Salim values-based attributes. The Asian benchmark of Singapore is a case that has used planning strategically in its quest for a technology-enabled intelligent development that is holistic and incorporates its unique blend of multidimensional Asian values and political economy characteristics (Corey 2004a). Malaysia too has drawn on its spiritual values and practical qualities in the pursuit of its own brand of intelligent development (Corey 2000). It remains for the leadership, the planners and the stakeholders of Gyeonggi province to invent their own particular and special planning scenarios for its intelligent-development future. Thereby, the people of Gyeonggi province, a suburban region in need of identity, community and embedded in the vast megalopolis of Seoul, in the words of Confucius, will have "something to stand" for and something to plan toward.

Scenario 6: Scenario Planning as a Means of Generating Development Alternatives – The Case of the Relocation of the Capital of the Republic of Korea

The Issue and the Problem

During his autumn 2002 election campaign for president of South Korea – i.e. the Republic of Korea – Roh Moo-hyun pledged that he would have a policy to relocate the country's administrative national capital function away from Seoul. He was elected president in December 2002. During the 2003–04 period, the debate of whether or not the national capital of the republic of Korea was to be relocated became intense. The issue came to a head in October 2004, when the country's Constitutional Court ruled that the plan to move the Republic's administrative center was unlawful. We became involved with the issue six months earlier, when we were asked by the Seoul Development Institute to prepare a paper on capital relocations around the globe. From these cases of the relocations of national capitals, we were asked to discuss the implications for Korea. This presentation was made before the International Symposium on the Capital Relocation on September 22, 2004 in Seoul (Corey September 22, 2004b).

The Design and Intention

The method employed for the study was designed to empower the principal stakeholders of the Korean capital relocation issue to be prepared initially to go beyond just saying "no" to the capital relocation. Rather, our position was that if indeed the capital was to be relocated, then it should be done thoughtfully, with widespread discussion and participation and by means of strategic planning that recognized that the Korean nation some day might be reunified.

> The intended motivation is to stimulate members of the audience to assume the responsibility to engage thoughtfully and creatively, the full range of future development needs of, and options for Korea, its regions and localities. Further, the intent here is to stimulate members of the audience noted above to *construct and debate alternative development scenarios*. For example, the relocation of a national capital or some of the functions of the capital is a means to an end; but *what is the end being sought*? In the South Korean media, it is said that "balanced national development" is the principal goal for moving the capital. Further, if *the* end indeed is balanced national development, then are there other means available that might more effectively meet this end?
> (Corey September 22, 2004b: 43)

The study was organized in two parts: part one focused on the nature of capitals and the implications for Korea (Corey September 22, 2004b). Part two sought to make a case for developing options and alternatives to President Roh's specific policy and the particular locational choice that resulted from that policy – i.e. to relocate the national administrative center to the Chungcheong provincial area situated about 160 kilometers south of Seoul (Corey September 22, 2004c).

Knowledge Exploration

The nature of capitals was discussed by defining a capital, by drawing from the literature to identify eight types of capital cities, including the "intelligent capital"; and the critical success factors for a capital were identified. Selected cases were listed of the relocation of national capitals and countries with "multiple capitals." Thirteen national capitals and other capitals were analyzed in order to interpret and elaborate on such critical factors of relationality as leadership, championing, compromise, regime change and party politics, planning versus plan-implementation, path dependence, culture, tradition, regionalism, neighborhood effects, continuity of policy and priority, financing, time frame and the need to re-focus agency missions over time, role of design quality, administrative and managerial skills, and generational perceptions, among other issues. The implications of these, and more than twenty additional lessons were raised for the Korean capital relocation case.

Planning Scenario Construction to Produce Options and Alternative Policies

It was not sufficient to discuss only the implications of the relocation of South Korea's national capital – implications alone do not go far enough. The implications must be translated into planned decisions and action. Having assessed many earlier national capital relocation cases, the lessons that were derived were helpful in informing the Korean capital relocation debate. However, in the end, it is the action decision, not the debate, that counts.

Part two of the study therefore sought to move the capital relocation discussion toward more informed decision making for the Republic of Korea today, and also toward the possible unified nation of Korea for tomorrow. The development of options is central to informed planning and informed public policy formulation. Choice among strategic alternatives can lead to more effective planning practice. The essence of the planning scenario was to explore alternate locations to the one that resulted from President Roh's electoral campaign-pledge approach. The planning-scenario exploration was driven explicitly by the requirement that there should be a relationship between capital relocation and the country's long-stated goal of balanced national development.

As a foreigner and outsider, there was little likelihood that we could provide comparable depth of insight and localized knowledge as effectively as knowledgeable Korean informants. Consequently, in an effort to ensure that a form of such local knowledge and expertise was represented within the context of conducting an early planning-scenario exercise, we explored a range of individual and heretofore unrelated prior partial regional-development solutions by different Korean experts. We organized the solutions by layers that were ordered from macro-scale strategies to micro-scaled ones. In all, eight elements of layered alternative future relationships were conceptually aligned, linked and integrated. This hierarchy of layers ranged from the macro-regional layer of transnational corridors that extended spatially beyond and through Korea, including areas that encompassed likely alternative locations for the relocation of the nation's capital. At the other end of the hierarchy were the operational actions that were required to plan, implement and schedule the actual development of a new capital city of Korea. The hierarchy that resulted from the scenario was composed of a global layer; a Northeast Asia layer; an all-peninsula layer; the two-Koreas national layers; and subnational layers.

The work of Professor Yu Woo-ik was one of the Korean expert informants selected. Yu developed the following criteria for relocating the capital from Seoul after re-unification (Yu January–February 1996). From this work, the new location for the capital should take into account:

- Korea's history.
- The pride and sentiment of both the South and North Korean peoples.

- Accessibility to the Asian continent.
- "The Greater Kyeonggi Bay Concept" which would function as the Peninsula's principal trade and communications window to Asia by steering the local development relationships among the following:
  - Incheon International Airport.
  - A new river port at the mouth of the Han river to serve both North and South Korea; it would function as a subsidiary to Incheon port and would be connected to Seoul by the construction of a canal.
  - The Kimpo-Kanghwa area might accommodate industrial estates and a teleport hosting state-of-the-art information and communications systems.

In the end, the planning scenario concluded that the best area for the location of a new capital for a reunified Korea would be somewhere between Seoul and Kaesong. This area meets Professor Yu's capital relocation criteria and the ancient Kaesong city area location taps into Korean history and culture in that it served as the capital of the Koryo Dynasty (AD 918–1392). Yu envisioned the specific location for the new capital being near Paechon, between Kaesong and Yonan. This location reinforces the Pyongyang–Seoul segment of the peninsula's structural axis (cf. Dublin–Belfast corridor); the new capital here can use the Incheon International Airport, and water from the Yesong River and the new port at the mouth of the Han river. Most importantly, from the perspective of congested Seoul, much of the traffic generated by this new all-peninsula capital to and from the airport and other Incheon area and west coast development would not have to pass through Seoul. This alternative vision has the significant benefit of both relocating capital functions from Seoul, but also keeping them in sufficient proximity to take advantage of Seoul's "engine of growth" value-added complementarities. Further, this location is geographically and psychologically accessible to all Koreans, both North and South.

Part Two of the Korean capital relocation planning study also sought to catalyze fresh thinking on the parts of Korean public policy and development-planning practitioners and their stakeholders toward the issue of the relocation of the national capital of the Republic of Korea. A reframing of the capital relocation issues was needed, along with a more in-depth consideration of the topic. From dominant past South Korean planning approaches, the growth poles and growth centers concept was applied in the country's regional development. However, in considering, for example, the full range of Plummer and Taylor's six reviews of the principal spatial theories of economic development (2001a; 2001b; 2003), five other bodies of theory await mining for application to the future reunified Korean case, especially within the context of the new changing dynamics of the global knowledge economy and network society. An important subtext here is that individual theory application must be empirically congruent with its conceptual intent. For example, growth pole concepts were formulated in an earlier era and in the context of natural resource-based and related economic activities. Within the Korean modern historic context, they have demonstrated variable performance when applied to industry-based and knowledge-based economic functions. Further, these various economic functions have their own spatial and locational patterns, scales and requirements; so, in planning for particular economic functions and conditions into the future, the appropriate geographic model in an appropriate projected empirical context must be used for the greatest planning effectiveness.

The earlier, Part One paper (Corey September 22, 2004b) also derived over twenty planning and strategic lessons from analyzing and comparing capitals and other related cases from elsewhere. Only a few of these lesson sets were mentioned, and then only superficially. The remainder of these lessons, plus the theories just noted, represents a rich source of conceptual material to help stimulate a wide range of strategic planning-scenario options for attaining greater balance in Korea's national development, including the issue of relocating the national capital.

In the end, the intended contribution of Part Two of the Korean capital relocation project was to illustrate that it is quite feasible to reframe alternative approaches to planning. In this case, the illustration operated from the premise that viable strategic alternative futures could be constructed from the stimulation of countering what was perceived to be President Roh's heavy-handed non-participatory approach and location decision for a new capital. By means of planning scenarios, we illustrated the construction of a synoptic strategic planning approach. It is an approach that accommodated the unique future development requirements of Seoul's city-region and of the country's requirements for balanced national development. If one takes a relational planning perspective to these particular sets of issues, Korea's unique assets and traditions indeed can be used to construct an even more globally competitive and happy nation.

This part of the capital relocation study therefore has been normative. It brought together prior thought by a number of Korean policy thinkers and planners about their individual solutions to Korean national capital relocation and balanced national development. The purpose in presenting and connecting these positions was to suggest that alternatives and resultant policy choices can, and should be part of an important national policy debate that ultimately ought to drive any large-scale public and private development project and investment.

South Korean strategists and developers have proven repeatedly that they have the technical and creative capacity to execute successfully such mega development efforts as the summer Olympics, a world's fair and a World Cup event characterized by innovative multiple venues and bi-national organization. No less success and excellence should be expected in planning for and implementing the possible relocation of the national capital, and/or addressing the complex issue of balanced national development, both for South Korea, and later for the time of possible reunification of the two Koreas. Korean society and culture have been under development for over five thousand years.

From this rich history and culture, Seoul has evolved as *the* capital of the nation for the last six hundred years plus. As a result, this sprawling and dynamic city-region has developed the only agglomeration of ingredients required to lead Korea into being an effective competitor in today's global knowledge economy and network society. Seoul's special position therefore demands special policy and planning attention.

In essence, what are the principal ingredients for pursuing the interdependent development visions of the national capital and balanced national development throughout the country? The lessons of relational planning practice suggest strongly that there are several imperatives for the times ahead in Korea. New forms of governance must be invented and practiced in the realms of public and private policy and development planning. These should be collaborative and partnering behaviors that must cross political-economic sectors at all levels – i.e. from top to bottom of society. In addition to generating layers of spatial strategies appropriate to their respective scales and substantive relationships for future states of planned development, there must be adherence to the tested principles and patterns that certain locations are congruent with and nurture some economic functions better than others – i.e. there are many different geographies. At local and regional levels, the highest priority for continuing attention and investment must be steered differentially as influenced by the most critical success factors for effective development in the global knowledge economy and network society – i.e. in developing quality human capital and in sustaining a creative enterprise culture.

Innovation and even more thoughtful effort must be expended in this case because a great deal is at stake. To paraphrase planner An-Jae Kim, as he looked to the expected future need to strategize and plan for a unified Korean peninsula, he wrote: "we the people living on the Korean Peninsula, should do our best to transmit a worthwhile land and settlements to our descendants through effectively unified development and conservation of only one country with unified wisdom and efforts" (Kim July 7–11, 1993: 52). The case of South Korea's capital relocation planning scenario as described here serves several purposes. Most importantly, it is an example of a tactic that may be used to produce alternative strategic planning solutions to future development. During the summer of 2004, the public discussion throughout the Republic of Korea revolved around President Roh's insistence on proceeding to relocate the administrative functions of the national government from Seoul and to the Chungcheong provincial area to the south. The framing of the discussion then was blurred as to the real reasons behind the relocation of the capital – i.e. several varying reasons were given over time. This transpired within the context that relocation of the national capital had been debated off and on for a generation. The public discussions of the summer of 2004 generally were not about alternatives to the president's scheme. The planning scenario approach discussed here represented an alternative approach and it sought to steer the needed public discourse in what were intended to be thoughtful and socially responsible directions.

Scenario 7: A Local European Application – The Isle of Wight, England

This planning-scenario case is a complement to Scenario 1, "Planning Scenario for Training Regional and Local Planners, East Central Michigan" (see pp. 140–4). Similar to the Michigan case, the Isle of Wight scenario also is regional and local in scale. However, it is an illustration of a more fully implemented planning process. Whereas the purpose of the Isle of Wight case was to conduct actual plan-making involving the community's various stakeholders, the purpose of the East Central Michigan regional case was to rehearse the Planning Support Team so that its members would be prepared to engage the local stakeholders in general, and especially the Regional Strategic Planning Steering Group of stakeholder representatives. Since the intent in the Michigan case was only to train other planners, including citizen planners, that planning process ended at the conclusion of that training and learning activity, whereas the Isle of Wight process continued and was completed through to plan implementation.

Stakeholders in England Engage in Action Learning for Planning

Marcus Grant has documented a local-scale project of self-described "loose scenario" planning that he facilitated (Grant December 2004). The processes that he introduced resulted in a strategic-planning intervention in tourism planning and policy development for the Isle of Wight. This participatory and capacity-building planning process began in 2000 and continues. The goal of the effort was to reorient tourism on the island to be more environmentally inclusive and to strengthen the social benefits of tourism there. Before the intervention, tourism policies had been driven principally by economic goals.

This case is relevant, particularly because it describes and examines the relationalities of working with individuals and stakeholders who are from various organizations, some of which are large and powerful, others of which are much less so. This is similar to several of the other planning-scenario cases reviewed above. Further, the case demonstrates the functions and behavior of the roles of different actors in a regional planned change process, such as a role for a local steering group. The case also illustrates the challenges of doing planning in a fragmented and complex industry. An important objective of this tourism intervention was to have developed a "strengthened capacity" for influencing future policies. Mindset change, collaboration and consensus building were found to be instrumental in transcending the diversity among the various actor-organizations. Each of these characteristics was part of the context of the other planning scenarios described in this part of the book.

This case study is noteworthy as an effective example of local-scale relational planning – i.e. reflective planning. Between phases of the project's planning process, time was built-in for reflection. The approach was congruent with Patsy Healey's "communicative turn" in particular (Healey 2002) and relational planning principles in general. Grant's process for the Isle of Wight relied on consensus building. As a result, social benefit and environmental inclusion were incorporated into the island's new tourism strategic planning. This is an example of a local planning intervention being instrumental in changing the planning mindset and behavior (Grant December 2004: 17). Grant's planning study provided evidence that a reorientation indeed had occurred:

> The reorientation required is to change from a view that 'tourism is a beneficial economic force in any form,' to a view that 'tourism's key role is as a driver for cultural and environmental betterment and local economic resilience.' This is a radical stance for the local tourism traditionalists. A tourism planning process that attempts to assist this form of tourism development needs to support stakeholders in such a reorientation. This study attempted to support a reorientation to the radical position through the use of multi-organizational working that incorporated techniques and tools more often found in the corporate change programmes.
>
> (Grant December 2004: 17)

The change-agent role and work of Marcus Grant for the empowerment of the Isle of Wight's diverse tourism stakeholders incorporated the following relational ingredients and actions; they are representative of the current state in the evolution and development of today's regional and local relational-planning best practices:

- These efforts do not occur overnight. It is important to have the time necessary to enable the convergence of the stakeholders' collective interests and the needed gestation of their core vision to emerge. The "completed" process in this case study occurred over several years. Because it frequently is the case that these processes need continuous energy and attention, completion may mean passing the effort on to other actors as, for example, it progresses from the action planning stage to the implementation stage.
- Sufficient investment in time can facilitate the development of the capacity necessary to enable the planned-change process to continue after the non-local change agent – i.e. Marcus Grant – has had to disengage from the project.
- Effective governance entails a multi-organizational and many individual-person arrangement that can evolve into an organization or organizational network that plans and/or implements plans. This is a networking approach that is congruent with the ICT-enabled environment today within which innovations must emerge and come to fruition. Use of this emergent and asymmetric-power organizational arrangement directed towards reorientation of economic, social and environmental functions, requires governance behavior that employs: collective processes that

> translate complex and fragmented activities into shared visions; coordinative approaches that convey a "collaborative advantage" in that the cooperating actors are able to realize their own goals better by working together rather than operating separately – this is "shared working for mutual gain" (Grant December 2004: 5); effective participatory techniques that sustain interest from actors with widely varying capacities; integration of multiple objectives that must be cast in relationality one to the other on a sustainable basis; a steering group is essential, along with the requirement that the stakeholders and wider community be informed and able to influence governance decisions; these arrangements raise issues of representativeness and therefore this requires definition and attention; transparency and accountability are essential in such extra-governmental private and public arrangements or partnerships; such governance arrangements are likely to be impermanent or temporary, this too requires special attention; these relationships have been acknowledged as stimulating futures planning, such that "their results are flexible and innovative, producing policies which are responsive to local needs" (Grant December 2004: 5).

Importantly, Grant's processes also incorporated equity issues. During the action phase of his action learning process, attention was paid to the involvement of underrepresented groups. This case study demonstrated that reliance on a multi-organizational network can have the benefit of contributing to equity. The inclusion of many stakeholder interests in the planning process can help to increase equity, as well as harmony and efficiency (Grant December 2004: 5). The importance of involving a wide range of participants and their diverse interests can contribute to the general state of equity within the context of the project.

Grant's efforts produced a planning and plan-implementation capacity on the parts of many of the representative stakeholders to be able to execute the planned initiatives beyond the initial planning phases. Indeed, the project's visioning extended 20 years into the future. Too often project planning gets limited to the short-run, and the immediate and pragmatic concerns of funding for the next budget year or so. Rarely are complex and multifaceted visions realized soon. Consequently, the criticality of taking and incorporating the long view was illustrated in this case study. This is particularly relevant, given the centrality of the planning that was focused on the sustainability of the environment.

Finally, this case study is a model for translational research in planning. Refer to Part V, section on Relational Planning Concepts, A–Z; see Translational Research, p. 219. The Isle of Wight project functioned at the seam of the world of planning practice and the world of planning research. The researchers for the project brought to bear the latest in corporate planned change techniques to the much less controlled environment of the community, and its highly variable and often divergent interests. Because the researchers facilitated an action research approach in which the stakeholders were participants in advancing the creative use of the knowledge that was introduced by the researchers, both the state of the art of relational planning practice and relational planning findings were strengthened. This carefully crafted case has set a high standard for best practice for today's local and regional planning within the context of implementing relational planning in the tourism sector.

Scenario 8: The Future of European Regions – A Continental Scale Scenario

At the transnational scale for most of the continent of Europe, an informative multiple scenario planning exercise was developed and described by Fink and Owen (Fink and Owen Spring 2004). These scenarios were designed to shed light on the future conditions of the European Union as a result of enlarging from 15 member countries to 25 members. The enlargement involved the addition of ten countries from Eastern and Central Europe. This scenario project was designed to explicate how the regions of Europe could develop over the next 20 years. The type of scenarios discussed by Fink and Owen are "succinct presentations of possible futures" (Fink and Owen Spring 2004: 9). In contrast, the scenarios discussed above – e.g. for East Central Michigan and the Isle of Wight – were principally planning scenarios, wherein future images are willed into reality by means of collective systematic decision-making, rather than putting forth informed forecasts of possible futures. The Fink and Owen case also functions appropriately on the increasingly critical level of the region – i.e. the subnational scale of areas that often manifest their own local blend of culture, social and economic dynamics. This scale is central to the core contribution of this book to support the needed changes to urban-scale and regional-scale planning practice for engaging the global knowledge economy and network society.

Even though the case study of scenarios for the future of Europe's regions is not an example of a planning scenario per se, it is referenced here because it provides detailed and useful insights into the methodology of conducting a scenario exercise that can result in a rich set of outcomes that can be valuable in planning policy responses in anticipation of some of the likely futures that were suggested by the scenario-development exercise. Further, this case reveals clearly the linkages and relationships among the many diverse, as well as similar functions and factors that are so critical to understanding today's network society. Also, the continental scale of this scenario exercise relates to the organizing theme used in this book of viewing the global knowledge economy initially through the lens of the three principal technology-economic regions of the world, including Western Europe. Simply described, scenarios of possibilities can be facilitative of formulating responsive planning scenarios.

In this case study, the intense three-day scenario-development exercise in the form of a conference was driven by three questions: (1) When/how will

Europe integrate? (2) Will European regional cultures survive? (3) Will Europe prosper? The conference participants generated 72 influence factors. Scenario software was used to reveal patterns of consistency among these many factors of foresight. "Projection bundles" were constructed. In turn, these were used to produce five scenarios of possible futures for Europe's regions. They were:

- Scenario I – Europe's Regions Roll Back in a (Re-)Nationalized Past.
- Scenario II – Regional Social Qualities Beat Global Growth Pressure.
- Scenario III – Strong European Regions in Transit Between Old Habits and New Customs.
- Scenario IV – United Europe of Regionally Networked Information Communities.
- Scenario V – United States of Europe in the Footsteps of the US.

Each of these scenarios was composed of various likely future characteristics. In turn, these were organized into categories of "who wins" and "who loses" by each of the five scenarios as each might become a reality. The scenario formulation process next produced a "roadmap." It identified the driving forces that could result in the five scenarios of projected trends of possible future development. These then were interpreted, with the scenario conference participants being asked to evaluate each scenario by its perceived highest and lowest probability of being realized in future. The final outcome was that the conferees foresaw as most likely, a Europe having developed into a continent of highly networked regions as envisaged in Scenario IV. Each scenario also was assessed by the forces that could stimulate the development of that particular scenario, and indicators were identified for use in monitoring the status of the trajectory for each scenario outcome.

The authors of this scenario-development case concluded that:

> The particular focus on Europe as a system of diverse regions, rather than just the traditional macro-political view of it as a continental whole, provided new insights and perspectives on the different ways in which these structures and communities could evolve in the years ahead.
>
> (Fink and Owen Spring 2004: 22)

This scenario process was seen as having credibility and usefulness because it produced surprises, paradoxes and conflicting characteristics within scenarios. Such inconsistencies seemed to the participants to reflect the real world of tensions and opportunities alike. The case also provided fresh insights into Europe's possible internal future, as well as its external future relations – for example, the likelihood of a new era in future relations between Europe and the US seemed high (cf. Moller 1995).

Policies Change

Ultimately, as these planning scenario activities evolve and become widespread throughout their respective city-regions, and when sufficient support has developed for local planned strategies, action may be taken to formulate policy changes. In time, new policies can lead to new programs and regulations to facilitate the new development approaches. These changes need to take the form of new public policies, institutional and business alliances, as well as many other transorganizational linkages that result in the reinforcement and strengthening of the voluntary networking that cumulatively can produce new governance for the region.

Surveys can be useful in stimulating processes of new policies formulation. For example, as a result of a series of statewide household telephone surveys that have been commissioned by the Planning Support Team in Michigan over several years, we have been able to list a number of policy recommendations. For a listing of some of these policies, refer to Frederick December/January 2005 and Wilson *et al.* 2004.

The British Government has commissioned an international benchmarking study of most effective e-economy policies (Booz Allen Hamilton November 19, 2002). Findings from such studies can be helpful in enabling local Regional Strategic Planning Steering Groups of stakeholders to formulate their own tailored set of new and changed policies. Such benchmarking studies can assist in setting both content and process quality controls for designing localized evaluation criteria. This e-economy benchmarking study included the following countries: Australia, Canada, France, Germany, Italy, Sweden, the United Kingdom and the United States. Examples of some of the world-class successful policies included fostering a strong market environment, such as successfully promoting cluster industries, including venture capital; promoting ICT in education; political leadership, which is associated with most of the benchmark countries, and those regulatory environments which are generally open to all; investment of public funds to close the urban–rural divide; best practice measures to enhance the extent and quality of national infrastructure; promoting citizen readiness, including tax reform to increase uptake of personal computers and improve citizen IT skills; fostering strong business readiness, including reducing the regulatory burden and brokering industry collaboration; fostering strong government readiness; effective policies driving government uptake; leveraging the role of the public sector as a player in the health industry; encouraging state and local government service delivery; user-centric service delivery, resulting in strong uptake; developing centralized online procurement for government departments; and, among other best practices, policies for online tax systems. The authors of the benchmarking project on the world's most effective policies for the e-economy stated that by identifying these good-practice policies, it should not be inferred "that these policies can be transplanted piecemeal between countries" (Booz Allen Hamilton November 19 2002: 164).

As policies planners and regional and local-scale stakeholders engage in formulating new, more responsive policies as a result of prior strategic planning, the following "good practices" should be used. According to Australian scholar Tony Sorensen, effective public policies should be characterized by: "(1) accurate problem [issue] definition; (2) reasonable understanding of problem cause; (3) clear allocation of responsibility; (4) technically competent (if not best practice) management strategies; and (5) effective monitoring of policy performance and policy revision in the light of assessment." He stated further that "research is an essential component of all five components" (Sorensen 2004: 1). Such research should be an independent assessment and draw on state-of-the-art theories and concepts. Rhonda Phillips has provided a useful guide for evaluating technology-based economic development programs and policies, including the use of survey methods (see Phillips 2003).

Continuous Learning

Having become *aware* and having acted on that awareness by implementing the remaining phases of the ALERT Model, it is critical to maintain the momentum of action. As the planners and their client stakeholders learn by doing, they will want to add to their stock of knowledge and learn even more. Developing learning communities may be pursued; refer to Part V, section on Relational Planning Concepts, A–Z; see Learning Regions and Innovative Milieux, pp. 208–9.

The ALERT Model calls for each of its phases to generate outcomes. Feedback from these results must be communicated to the appropriate phase in a systematic, iterative and recursive manner. Simultaneous with this process of plan-implementation, continuous feedback, replanning, monitoring and evaluation of this process of planned change should be enriched further by continuous learning. A simple, but rewarding tactic is to form community book clubs. They would meet on a regular schedule and operate from readings and books that instill new substantive knowledge and, in the process, provide ongoing stimulation and interest in the continuing implementation of the ALERT Model process. The kinds of books that should be considered for book club reading assignments and discussion might include, in no particular order, such selections as:

- *Dark Age Ahead* by Jane Jacobs (2005);
- *The World Is Flat: A Brief History of the Twenty-First Century* by Tom Friedman (2005a);
- *Collapse: How Societies Choose to Fail or Succeed* by Jared Diamond (2005);
- *The Past and Future of America's Economy: Long Waves of Innovation that Power Cycles of Growth* by Robert Atkinson (2005);
- *Tipping Point* by Malcolm Gladwell (2000);

- *Blink* by Malcolm Gladwell (2005);
- *Moyers on America: A Journalist and His Times* by Bill Moyers (2005);
- *Confucius Lives Next Door: What Living in the East Teaches Us About Living in the West* by T.R. Reid (1999);
- *The United States of Europe: The New Superpower and the End of American Supremacy* by T.R. Reid (2004);
- *The European Dream: How Europe's Vision of the Future is Quietly Eclipsing the American Dream* by J. Rifkin (2004);
- *City A–Z* by S. Pile and N. Thrift (2000).

The full citations are included in the References section.

In addition to organizing the book clubs, the reading assignments and discussions, the local planners and stakeholders also should consider other divisions of labor so as to be able to maintain ongoing *awareness*. For instance, some participants in the ALERT Model's planning process might volunteer to monitor generic progress and change in such relevant domains as economic development, ICT, science and technology, and development planning and policy issues. A valuable source of up-to-date economic development news is the *SSTI Weekly Digest* that is e-mailed from the State Science and Technology Institute (State Science and Technology Institute n.d.). Another useful source is *Brookings Metropolitan Policy Program*; by joining the listserv of this program, weekly e-mails can be received with news and linkages to web-based publications on policies affecting metropolitan regions (Brookings Metropolitan Policy Program n.d.). Since California often has been a bell-wether for innovation in many domains, policy research emanating from the Public Policy Institute California (PPIC) merits monitoring and study; one may subscribe online to the monthly news from PPIC (Public Policy Institute California n.d.). These newsletters, readings and suggestions are illustrative. The planners and their stakeholders also can generate their own tasks and resources to maintain momentum and continue to learn. Analogous to business-plan competitions, another useful tactic, both to maintain *awareness* and to stir up excitement and creativity for innovation, leaders in the local region may consider sponsoring annual intelligent development plans for the region. Awards and prizes should go to the winners, and special outreach efforts should be made to ensure that submittals come from local schools and nearby universities, as well as from business, government and other individuals. It is important that the community's full range of generations be represented and participating in such continuing *awareness* activities.

Practicing Relational Planning: By Practitioner-planners and Academic Practitioner-planners

Most of the discussion of the book has been directed at community-based practitioner-planners. This is in recognition that there is an urgent need to hasten changes in the current practice of planning to reflect better the new and constantly changing realities of the global knowledge economy and network society. However, it also is important to stimulate academic and university-based planning practitioners to change their practice. Through their students and research, planning academics have an important role to play in contributing to the transformation of the planning profession as has been advocated by Graham and Healey, Sanyal, Markusen and Brooks (see Part II, p. 40). This transformation should result in planning practice that is compatible with the new and dynamic relationalities of the global knowledge economy and network society.

The primary activity of academic planning practitioners is scholarship. This takes the form of research, publication, teaching and service. In order to contribute to the transformation of the profession, these scholarly and academic planning practice behaviors need to be reoriented to be more "relational," in the sense that has been called for by Graham and Healey (October 1999). For example, by orienting research and teaching toward relational planning practice, we can begin to create a body of experience and knowledge that will enable planners to be more significant and strategic contributors to today's regional and local development policy formulation and plan-making. The outcomes of planning and development scholarship are consumed primarily by academic peers, by pre-service students, and by non-academic peers – i.e. planning practitioners working in the community and engaged in continuing professional development. It is through these clients that we can help move the planning profession into more central and development-relevant roles in the new environment of the global knowledge economy and network society.

We have begun this process by steering some of our own scholarship in this direction. For example, in order to obtain feedback on the construction of the Model, the concepts underpinning the ALERT Model have been presented to various kinds of practice audiences; some of these included: (1) regional-level communities of general citizens, local government officials, business persons, nonprofit organization members, educational institutional representatives, local planners, extension agents, and so on, meeting in localities across Michigan; and (2) a large professional planner audience at the American Planning Association's 2004 National Planning Conference in Washington, DC; this presentation was made as part of the new conference track entitled, "*Research that Supports Planning*." Wilson and Corey also applied the principles underpinning the ALERT Model to several cases in Asia – i.e. in Southeast Asia – and to Korea (Corey and Wilson 2005b; Corey and Wilson 2006; Corey December 19, 2003; and Corey 2004b and 2004c).

Central Theme of the Book

A key part of the core message of the book is that members of the planning profession need to change their mindset and practice behavior to be congruent with the dynamics and uncertainties of the global knowledge economy and network society. Both academic planning practitioners and community-based planning practitioners need to take into account and act on these dynamics, for they require the utilization and practice of new relations and different strategies and tactics. The principal question is how might planners go about transforming the practice of the profession into a more effective, relevant and contributing local public-policy stakeholder in today's new economy and society?

The Challenge of Scholarship, and Constructing and Offering Relational Planning Education Opportunities

We call for planning scholars to devote their research, publication and service effort to the advancement of the practice of relational planning. In recent years, there seems to be something of a renewed and growing interest in the practice of planning. Among other examples, this has taken the form of (1) a track at APA's National Planning Conference on research that supports planning, as just noted above; (2) the American Institute of Certified Planners has initiated a quarterly online publication that is posted on the APA website (*Practicing Planner*); and (3) a new international peer-reviewed journal has been established as an important publication outlet for innovation in planning practice and scholarship: *Planning Theory & Practice*, edited by professors Patsy Healey, Leonie Sandercock and Heather Campbell.

Planning academic practitioners working in higher education institutions have the additional challenge of ensuring that they are offering both pre-service and in-service professional educational opportunities to their students and non-academic practitioner colleagues – i.e. for continuing professional development. Given the technology-enabled environment within which relational planning may be practiced, courses and programs in Geographic Information Science (GIS) can be an initial stimulant for such curriculum reform. However, this beginning is insufficient. It would be important also to provide societal, economic, cultural and policy context, such that GIS techniques are learned within the full range of content systems and ICTs that planning practice engages. Thus complementary courses in ICT-enabled planned development theory, method, philosophy and practice should provide the needed context for technical learnings such as in GIS. It is important that our educational and scholarly practice reflect the new realities of the global knowledge economy and network society. Thereby, the seeds of changed planning practice, mindsets and governance behaviors may be learned and tested by means of supervised pre-professional experiential learning.

Planning Education

In some parts of the world, ICT-facilitated regional planning – i.e. at the substate scale – has evolved and matured to the stage where practitioner-planners no longer can be content with planning digital development – i.e. the provision of access to ICT infrastructure. While it is essential to plan for and provide access to the full range of ICTs today, merely increasing the availability and access of cyber infrastructure in localities is insufficient to the demands of today's and tomorrow's global competitiveness. Practicing planners also should be strategizing and planning for content and creative application that takes advantage of the new development opportunities that are enabled by ICTs, and science and technology-stimulated economic development. Because some planning practitioners and planning practitioner-scholars respectively have engaged in and researched digital development, a formative, but potentially useful collection of relational theory and concepts has emerged – (e.g. Graham and Marvin 1996; Graham and Marvin 2001; Graham 2002). Some of this work was selected, aligned and incorporated here so as to be able to suggest practical relational planning approaches for the formulation of strategies and tactics that are congruent with and supportive of the more holistic and comprehensive development that should be responsive to and enhancing of the complexities and power relations of the global knowledge economy and network society.

With these and other emerging relational conceptualizations, today's regional planning practitioners may more effectively move beyond digital development which is focused on ICT infrastructure access to the content of intelligent development that is informed by appropriate theory – i.e. relational theory, sound empirical analyses – and thereby realize more effective contemporary planning practice and development outcomes.

The most effective regional and local planning practitioners and planning practitioner-scholars will continue to learn from experimentation and innovative practice with the new intelligent development opportunities that are offered by addressing the full range of these and other relational approaches. In future, promising additions to this emerging body of relational theory-informing planning practice may include the integration of actor network theory; more attention to the division of labor by phase of the planning process; the elaboration of actors, roles and agency inherent in the Program Planning Model and its organizational relations; the Program Management structure for the purposes of planning and innovation generation; and equity planning theories, among other concepts and theoretical constructs. Further, more planning scenarios for each of the six Plummer and Taylor-reviewed locational and local economic development concepts and theories (Plummer and Taylor 2001a) need to be constructed for representative regions and localities. In addition to the need for planning practitioners to engage and practice such relational approaches, it is believed that planning theory also can benefit by having planning theorists actively take up the many challenges of advancing relational approaches as part of the general body of developing planning theory. This task might be accomplished most effectively by planning theorists and planning practitioners working closely together, and with the common goal of taking the profession more centrally into the local development of a country's regions and communities. In this context, translational research should be an explicit shared value of the planning practitioner and planning practitioner-scholar alike. Refer to Part V, see section on Relational Planning Concepts, A–Z; see Translational Research, p. 219. Reinforcing this need, planning scholars Dowell Myers and Tridib Banerjee have written:

> we must reconcile the chasm between academia and the profession by working together. First, we must recognize the division of labor among key sectors in planning: the academic field, the profession, and the practice. Academics can explore problem areas not yet codified into professional practice, and they can bolster the reputation and standing of the profession through their publications and interactions with other academic fields. . . . The profession must have a forward-looking position that anticipates this rapidly evolving future, prepares itself for future opportunities and challenges, and remains intellectually nimble enough to adapt to new professional and pedagogic challenges.
>
> (Myers and Banerjee Spring 2005: 128)

These formative approaches and suggestions are offered to the leaders, stakeholders and planners of regions and localities to consider as they continue to grapple with the complex forces of the global knowledge economy and such related questions as: (1) regional identity, (2) the need for planners to earn the trust of the region's stakeholders and for professional planners to demonstrate that they have needed and essential value-added services to provide, and (3) how ICTs and science and technology can be used in development planning to reinforce a sense of community and enhance new good-governance behaviors. Benchmark cases of such contemporary development and planning should be studied for their best practices. Cases as from the Nordic countries and

Singapore are suggested because those political economies have attempted to answer similar holistic and relational planning questions. Additional comparative case research is needed also on the spatial planning innovations underway in Europe. The European Spatial Development Perspective (ESDP) is a program intended to achieve balanced and sustained development of territories of the European Union (Faludi and Waterhout 2002). Experience from the results of the ESDP can be expected to generate lessons for improved spatial planning practice in other regions throughout the world. British planning scholar Patsy Healey has documented, compared and assessed these results for several regions across Western Europe (Healey March 2004a; Albrechts *et al.* Spring 2003; Healey September 11, 2003c). Her 2006 (forthcoming) book uses these cases to derive and advance the current body of relational planning theory. The American planning profession also is fortunate to have a scholar who has thoughtfully researched and published on the practice of planning over a long period of time; Howell Baum's body of work on this topic should be consulted as an important support for advancing scholarship, having to do with improving, keeping current and making relevant the practice of planning (Baum 2000; 1996; 1986). In order to integrate old and new planning practice and reflection (Schön 1983), the results from more cases of relational planning applications need to be researched, translated and incorporated explicitly into pre-service and professional development planning education.

The Practice in Planning Education Committee of the Association of Collegiate Schools of Planning (ACSP) issued a draft Planning Practice Report (Practice in Planning Education Committee July 23, 2004). The committee reported on its examination of how ACSP member programs engage planning practice. The ACSP Committee Report identified a number of specific activities intended to enhance the collaboration and interaction among both on-campus and off-campus practicing planners. The results of the work of the committee can reinforce the call of this book for academic planning practitioners and community-based planning practitioners to use relational planning principles in their practice of planning.

For the improvement of the profession and its clients, university faculty planners are encouraged to steer some scholarly effort to the needed transformation of the profession and to experiment with and practice relational planning approaches, and in the process, to go beyond the formative suggestions contained here. Collectively, as we accumulate new relational planning practice experience and communicate the outcomes and lessons, planners and their client stakeholders in the world's regions and localities will be better prepared, through planning, to create a higher quality of life that contributes to the full development of the community's people and its institutions. This outcome will be the ultimate test of the needed new practice of planning to meet the various challenges and reformations of planning practice as articulated in Part II by Rodwin and Sanyal (2000), and Graham and Healey (1999). In the end, however, it is up to the people of the region and its localities to innovate continuously and to plan for the area to be one whose quality of life and productivity are interdependent with today's global dynamics and development opportunities, and to ensure ongoing development attention so as to realize increasing equity among a region's communities.

Toward a Global Planning Practice Learning Community

Given the rich local experimentation and development-planning innovations underway across the world's three major technology-economic regions, a compelling case can be made for systematic ongoing discussion, exchange of lessons and debate among planners, scholars and other development policy stakeholders of these regions. Collective and collaborative leadership from APSA (Asian Planning Schools Association), ACSP (Association of Collegiate Schools of Planning) and AESOP (Association of European Schools of Planning), is in order to ensure that learning and systematic sharing of successful and unsuccessful relational planning practices are available to all parties of interest – regardless of language and location (Kunzmann July 3, 2004). This might mean, for example, that these regional associations might focus significant joint effort to develop internally and inter-regionally discussion networks and journals that place high priority on sharing and "publishing" via various media, selected high-quality intelligent-development planning lessons and cases that are translated in whole and in part, and that explicitly draw on the particularities of innovative local planning experience. These initiatives should be piloted, their feasibility demonstrated, and profitable alliances explored with commercial publishers, university presses and libraries, and electronic media. In turn, these regional and local information resources would be made widely available and accessible to planning and policy practitioners and scholars around the globe. While all regional development is local and specific to that place, its people and institutions, these cases do have some comparative and analogous elements and results that can inform strategy and stimulate fresh planning elsewhere. For elaboration, refer to Corey and Wilson (September 2005c).

The Profile of the Relational Planner-practitioner

By way of capping this discussion, we should ask, "What does the relational planner-practitioner look like?" This profile applies to the professional planner and the citizen-volunteer planner alike. The principal defining characteristic of the effective contemporary practicing relational planner is one of a new mindset. This perspective is one of viewing the current and future world through the filter of multiple linkages and dynamic, often-blended non-linear relationships. The relational-planning mindset also sees the world from a futures perspective. In planning for the future, the relational planner relies on concepts and theory, especially tested bodies of theory for inventing effective and creative approaches to stakeholder-informed strategy formulation and plan development. The effective relational planning approach is based on new governance relationships that are multiple-actor in nature and representative of the major stakeholders of the region's localities. It is the relational planner's leadership responsibility to demonstrate relational planning behavior to and with the regional stakeholders, such that they too practice and routinize the new relational planning style. Over time, the behavioral objective is to embed relational planning increasingly into the way that planned development is carried out in the region. With focus and practice, this means that there should be an incremental shift from a dominance of old segmented perspectives and fragmented governance to a state of predominant relational integrated perspectives and highly coordinated policies and strategic interventions. An ideal end-state for relational planning practice would see a complete adoption of the relational principles, thinking and style that have been illustrated here in this book. A last element of the profile of the relational planner must include the reminder that even with full engagement and perfection of effective relational planning practice, these are not guarantees for success. The external economic dynamics and the locality's position relative to other places are in continuous flux. As a result, one should expect ups and downs in all of the measures of development.

The Successful Relational Planning Practitioner Will Use the Five Fundamental Relational Planning Lessons and Evaluate the Outcomes of the ALERT Model Process

As the practicing planner pursues the process of planned change inherent in the ALERT Model, it is important to monitor progress on the implementation of the content goals that will have evolved from the planning. It also is critical to monitor the status of the community's mindset change and improvement of governance among the region's stakeholders. Similarly, the investments in, and resultant outcomes of human capital development, the development of an enterprise culture, and the reduction of economic disparities across the region all need to be assessed regularly.

To stay abreast of changes in the mindset of the region's planners and their stakeholders, periodic surveys should be taken. One of the primary objectives for these surveys is both to derive information from the planners themselves for supporting regional and local relational planning in their planning practice within the explicit context of the global knowledge economy and network society. As noted above, it is essential to measure changes in the region by the five critical planning lessons. These are: (1) the development of human capital throughout the human life span – i.e. from early childhood development through to the formal primary, secondary and tertiary education and in-service job development education and training; (2) to achieve an enterprise culture in the region that hosts and supports entrepreneurship and innovation by means of cross-sectoral collaboration and partnerships among the business and nonprofit institutional communities; (3) the need to change the mindset of regional and local planners so that their practice behavior is congruent with the complexities and uncertainties of the global knowledge economy and network society; (4) the enterprise culture can be created and sustained by means of new governance behaviors that transcend the major stakeholder interests of the region and seek to ameliorate the fragmentation of the various governments, firms and institutions as they might converge and coordinate at the local level; and (5) explicit and activist attention to the region's economically distressed communities; this means identifying such disparities, their spatial distribution and developing strategies and plans to attain more equitable distribution for opportunities to engage the potentialities of the global knowledge economy and network society. These five elements are as diverse as apples and oranges, but they do represent needed outcomes and planning practices that are prerequisites to development success within the context of the global knowledge economy and network society. Consequently, it is believed that it would be useful to obtain empirical information that enables one to compare and contrast this five-lesson vision of improved planning practice to current planning practice, and then to identify what the planners perceive their practice needs to be so that research and support might be provided for their continuing professional development (cf. Breuckman April 2003; Singh July 2003).

Applying Relational Planning to a Nonrelational Planning World

One of the major challenges facing planners and their stakeholder constituents is that data and information used in strategizing and planning, when available, is in discrete, nonrelational forms. For examples of such local and regional-scale uses, refer to Gandhi *et al.* June 2005; LaMore *et al.* April 21, 2005; LaMore, *et al.* July 2004. For larger scale uses, at the state and metropolitan scales, refer to Atkinson November 1998; July 1999; June 2002; Atkinson and Gottlieb April 2001. These factual and spatial representations enhance our collective ability to conceptualize relational visions and construct realistic planning scenarios. One is reminded, however, that such challenges are a normal part of planning practice. Planners always must work under data constraints and other challenges, such as the pressures of time and tight deadlines, and less than perfect information. In the end, one must do one's best with what one has.

Some Final Cautions

It must be clear to the practicing relational regional and urban planner that fully grasping the development possibilities of the global knowledge economy and network society is not a silver bullet for successful development. One should be reminded that some of the most successful and well-managed localities still have economic difficulties. A few such examples include Silicon Valley, California; Portland, Oregon; Grand Rapids, Michigan; the European Union; and Singapore. Over time, each of these cases has demonstrated considerable success and knowledge-economy capacity in taking advantage of technology and science-enabled development opportunities. Yet each of these and many other such successful city-regions have experienced economic downturns. So in addition to having a modern foundation of digital development or ICT infrastructure, and a strong performance in the content of intelligent development, the constantly changing dynamics for even the most competitive city-region requires it to be resilient. The ability to reinvent its economic advantage and to sustain the dynamism of its society, quality of life and culture are key to further intelligent economic development in future.

Electoral politics too can play a role in maintaining a successful competitive edge for localities (Atkinson 2005). At all levels of government, there is a tendency to change or shift policies when regimes, political parties and elected leaders change. This can have a positive impact when new policies result in new high-wage jobs, for instance, but new policies can have the opposite effect. For example, a new governor of a state may feel compelled to shift development direction, thereby diluting the focus of the ICT and science and technology programs, and investments of the prior administration. A certain constancy of policies and investments – both public and private – are critical to continuous success in intelligent development. The new governance structures should pay particular concern to ensure that investment priorities are maintained in the driving factors of human capital capacity-building and the sustainability of the city-region's enterprise culture. Short-term expediency, at the expense of long-term development enhancement should be avoided. For example, cuts to educational budgets typically fall too heavily on such supportive and reinforcing knowledge-economy programs as art, music and performance; these fields enhance creativity and, in the long-run, combined with science and technology priorities, all contribute to the creation and maintenance of a local culture of innovation and improvisation so critical to competing among the forces of uncertainty of today's economic and societal environment.

Around the globe, we are living and working in local and regional environments where the governments that are there to serve us too often are indifferent, or not responsive to the development needs and inequities of their jurisdictions. Further, we as citizens, planners, economic developers and stake-holders in general have left the business of community leadership and governance to others. Too often also there is a tendency to feel overwhelmed by the big forces – i.e. of globalization, for instance – that impact us each day. Once you make the effort to commit to make yourself more aware of the issues and opportunities of today's rapidly changing economy and society – at the local level – you have taken the first steps toward engaging in, and contributing to the revitalization of our civil society. This volume has sought to support your interest in becoming more engaged and more effectively engaged. Now that some relational-planning development pathways from the three technology-economic regions from around the world have been illus-trated, the next steps are up to you to join with other like-minded citizens and plan to invent your community's new future and your locality's specific rela-tional pathway to that future. You have our best wishes for a creative and productive journey.

Epilogue: New Opportunities and Challenges for Planning Improvement

Planning is often taken for granted. It is easy for the public to assume that "someone" is taking care of the needs for visioning, organizing and managing the built environment and community well-being. With a century of practice and many positive experiences with effective planning in the past, the public should see the planning process as a fundamental element of urban and regional life. At the same time, the process of planning is often invisible, even if the end results are not. This is especially relevant for planning information and communication technologies applications, which Stephen Graham and Simon Marvin (1996) note as a challenge because the elements of ICT are often intangible and absent from the landscape. The invisibility of the process carries with it a need for trust by the public in planning as an institution, and public vigilance that community needs are being addressed.

The opportunities for learning and practicing the ALERT Model surround us – both locally and elsewhere. For example, the impacts of several major 2005 hurricanes on the city-region of New Orleans and the coastal region of the Gulf of Mexico has revealed a great deal about the criticality of local and regional planning to the economy and society. The lessons from these events are informative for the culture of planning in the US and the other regions of the world. As broadcast around the globe, these dramatic events have placed the need for visionary planning and effective plan execution center-stage in our awareness. Because of the focus of the globe and the scope of the re-development budget being committed to rectify these governance and planning omissions, this is a clear and rare opportunity for all of us to learn and practice most of the planning behaviors that have been developed and advocated throughout this book.

In addition to being able to practice the principles of the ALERT Model, thoughtful commentators on the New Orleans and US Gulf coast have provided us with complementary and reinforcing suggestions that are intended to stimulate regional and local leadership, problems engagement and planned action.

For example, the following three columnists have made insightful observations encouraging good planning practice and effective plan execution that merit our attention and mobilization for action: Neal Peirce, E.J. Dionne Jr and Thomas Friedman. Peirce has noted corporate executive leader Robert Grow's call for "assembling around the table" some of the best minds as a visioning team to address the redevelopment of New Orleans and the Gulf Coast region:

> All the governments would be there . . . but also people from America's highly skilled non-profit sector – organizations like the American Planning Association and the American Institute of Architects, the Urban Land Institute and the Alliance for Regional Stewardship, independent and noted experts in hydrology and flood control, transportation and housing, builders, insurers, mortgage bankers, others.
>
> (Peirce September 20, 2005)

As a result of these hurricane disasters, Congressman Earl Blumenauer of the Portland, Oregon area is quoted as saying that this is "the closest we'll ever get to a blank canvas" for truly thoughtful and intelligent planning (Peirce September 20, 2005). A process is advocated that produces scenarios for public debate, clear choices and actions. Using selected key words of, and paraphrasing from Neal Peirce, such a process should be open, democratic and not rigged, and it would have the following characteristics:

- Encompass the entire multi-state region.
- Create a sustainable region consisting of reshaped city-regions.
- Engage whole communities openly wherein the dispossessed could engage and have dialogue with the broader society.
- Use contemporary techniques such as electronic town meetings, geographic information systems and visualization approaches that depict how communities would be changed by various planned actions.
- Governments at all levels would behave as partners by listening and learning of the needs of the communities.
- Difficult, but thoughtful decisions would result from buy-in by citizens of the affected communities; this might involve decisions such as "focus on more compact development, mixed-use and mixed-income towns."

In addition to the engagement of citizens and institutions that are directly impacted, involvement from the private sector, the nonprofit sector and all the layers of government working on and from a shared vision is essential to a successful planning and implementation process.

In an analogous vein, E.J. Dionne Jr has called for the rebuilding of New Orleans as:

> a model for a better kind of city and the rebuilding of the Gulf Coast a model for a better approach to governing. The people of the region, not the lobbyists, need to lead in creating an environmentally sustainable, socially just and economically viable region.
>
> (Dionne Jr September 20, 2005)

E.J. Dionne Jr, like Pierce, draws inspiration from Congressman Earl Blumenauer who returned from visiting the Asian tsunami areas eight months before hurricane Katrina hit the US Gulf coast and asked on the floor of the US House of Representatives what would happen to New Orleans if a hurricane hit the low-lying location of the city-region. He has called for a revisiting of "past policies that encouraged development in dangerous places" (Dionne September 20, 2005).

Being located in the Southeast Asian city-state of Singapore at the time when hurricane Katrina hit New Orleans and the Gulf coast, columnist Thomas Friedman stated, "if you had to choose anywhere in Asia you would want to be caught in a typhoon, it would be Singapore" (Friedman September 15, 2005). His point was that Singapore places high value on good governance and leadership. As a result, the culture there for government is one of high expectations for competence, incorruptibility and a significant degree of accountability.

When one reflects on the various mix of characteristics identified by these observers, most of the basic ingredients for good planning and effective plan execution have been identified. The challenge before us now is to use the new awareness offered to us collectively as a result of the poor governance and planning performance in New Orleans and the Gulf coast as a stimulus to develop and sustain a culture of responsible planning behavior and improved civil society. Can we meet this challenge?

Part V
Support

Relational Planning Concepts, A–Z

Action Research

It is research that is shared by researchers and stakeholders. It involves the researcher acting as a change agent with community participants functioning as co-researchers (Grant December 2004: 5). Action research can be critical in developing working relationships and shared interests among a region's stakeholder representatives. In this context, it is research that is developmental and innovative, and its goal is to produce new visions and approaches to realize those visions by means of collective, participatory and multiple stakeholder investments of learning and planning accordingly (Clark 1972). Action research can facilitate the incorporation of local knowledge and expertise into the plan making for the region.

Actionable Knowledge

Actionable research tells one what is relevant; actionable knowledge "tells you how to implement it in the world of everyday practice" (Argyris and Schön 1974: xviii). For insights into the roots of the organizational change and organizational development tradition, see Marrow 1969.

Agency

Agency is an entity through which power is exercised – e.g. the board of trustees communicated its new policy through the agency of the school's headmaster or principal.

Alignment

In the context of the book's discussion, alignment stands principally for the lining up of knowledge-economy development policies and programs from the locality, to the region, to the province or state, and to the policies at the national-level. For example, in the case of the economies of the European Union, it also means that there would be alignment and harmonization of regional programs with continental-wide policies. Such alignment is indicative of congruence and agreement up and down the layers or levels of a technology knowledge- and science-enabled policy hierarchy. Alignment is a lack of fragmentation. A lack of policy fragmentation is intended to create a more operational, effective and pervasive policy environment for the advancement of knowledge-economy programs. For an example of the alignment needed for effective strategic planning, see the section on Good Governance, p. 201.

Benchmarking

The purpose of benchmarking is to improve the planning of the region and its communities by learning from other areas. It involves the regular and routine comparison of planning performance with the best such planning and planning organizations elsewhere. Benchmarking enables the identification of gaps in performance. In turn, these findings can be used to continuously improve the process and outcomes of the planning (Holzheimer Winter 2005).

Best Practices

These are ways that organizations – i.e. governments, companies and institutions – operate to lead to an intended result. Such practice is based on research and experience. A best practice has a proven record of success in improving performance and quality. Ideally, best practices are in the form of measurable factors; this facilitates peer systematic comparison and evaluation.

Biosciences

The following is a useful operational definition of biosciences:

> In addition to comprehensive nationwide studies of the industry, many states and regions have prepared analyses examining the local concentration of biotechnology-related economic activity. In almost all cases the definition of biotechnology is tailored to local perceptions. Every community, it seems, defines its local biotech industry in its own fashion, including and excluding sectors based on differing judgments. Almost all of these definitions include biotechnology as defined earlier, but they also

> reach out to draw in other activities under a wide array of terms including "biosciences," "life sciences," "biomedical sciences" and "healthcare technology." Many of these state and regional studies are used for marketing and promotional purposes; comparisons among local studies tend to be difficult or impossible.
>
> (Cortright and Mayer June 2002: 39)

> Biotechnology has been too narrowly defined. Broader definitions are needed to address the key components of a cluster: supplier chains; service referrals; geographical proximity; innovation; and collaboration connectivity. Traditional definition ignores [the] concept of cluster. It is no wonder the so-called experts can find less than a handful of "bio" clusters in the US.
>
> (Horowitz October 1 2002: 11)

Mitchell Horowitz of the Battelle Memorial Institute has defined the "biosciences" as composed of five subsectors:

1 Organic and agricultural chemicals (product-oriented)
   - Industrial inorganic chemicals
   - Fertilizers
   - Other agricultural chemicals

2 Drugs and pharmaceuticals (product-oriented)
   - Medicinals and botanicals
   - Pharmaceutical preparations
   - Diagnostic substances
   - Biological products

3 Medical devices and instruments (product-oriented)
   - Laboratory apparatus and furniture
   - Surgical, medical, dental, and analytical instruments and equipment [e.g.] X-ray and electromedical equipment

4 Hospitals and laboratories (service-oriented)
   - General medical and surgical hospitals
   - Psychiatric hospitals
   - Specialty hospitals
   - Medical and dental laboratories

5 Research and testing (service-oriented)
   - Biological research
   - Medical research
   - Food and seed testing laboratories
   - Veterinary testing laboratories

(Horowitz October 1, 2002: 12)

Biotechnology Industry

Cortright and Mayer (2002) have defined biotechnology in the following way:

> Biotechnology is a new and rapidly changing industry that has yet to find a neat, separate categorization in either the old Standard Industrial Classification (SIC) code or the new North American Industry Classification System (NAICS). Even so, a general consensus about the contours of the biotechnology industry has emerged from industry participants, investors, and a range of comprehensive studies of the industry. Rather than rely on secondary statistics compiled by government agencies in broad industry classifications, industry analysts and researchers have relied heavily on primary microdata – firm level statistics on employment, investment, and activity. The present study follows this generally accepted biotechnology definition that has emerged, and it employs microdata from a variety of sources.
>
> (Cortright and Mayer June 2002: 36)

Broadband

"A reference to high-speed data transport. The Federal Communications Commission defines broadband as 200,000 bits per second (200k) or greater." The Rural Internet Access Authority (RIAA) of North Carolina also has offered a comprehensive definition of the elusive phenomenon of "broadband." The RIAA has operationally defined broadband to

> provide the necessary bandwidth access to support two-way quality Internet services to the home and businesses. This requires at least 128 Kbps for residential subscribers and 256 Kbps for business, but avoids a 'one size fits all solution' so that rural communities can fully implement the economic and social benefits of Internet applications as defined by RIAA initiatives. Broadband access should be forward looking and embrace innovative evolution in technology and applications.

The authority's definition of broadband goes on to articulate eight additional principles, including: (1) provide incentives/subsidies/tax credits if broadband is not otherwise provided; (2) technology neutral, this "recognize[s] that all technologies have an opportunity to be used in broadband deployment"; (3) ensure community awareness of the benefits of broadband Internet through outreach and education; (4) involve [state] North Carolina government; (5) promote leadership participation by the vendor community; (6) encourage local involvement through the e-communities initiative recognizing that it is critical to broadband deployment – i.e. "involve communities through outreach and education to ensure that affordable access to and knowledge of how to use the

Internet is available to all citizens"; (7) provide rural Telecenters – i.e. "implement Telecenters in rural regions of North Carolina to facilitate citizens and businesses on Internet training, broadband access and business development"; and (8) recognize the relevance of federal involvement – i.e. "ensure that attention is paid to ongoing Federal legislation, regulation and assistance as this is relevant to the provision of equitable quality broadband access in rural North Carolina" (Rural Internet Access Authority January 2002). The authority now is called North Carolina's e-NC Authority.

Charrette

A charrette is an intensive effort to complete a task or a project within a limited time period. For use in planning, this activity originated in nineteenth-century Paris to describe the practice of architectural students completing their projects at the eleventh hour and rushing their models and drawings to the Ecole Des Beaux-Arts on a cart, or "charrette." Modern planners have employed this design tactic to facilitate the production of plans rapidly and collaboratively.

Cities of Business Leadership and Enterprise

Compare and contrast this concept with O'Mara's "cities of knowledge" concept below. Cities of business leadership and enterprise are city-regions, the high-tech origins of which are attributable to initiatives and actions by the business community. For example, early economic development of the area of Kitchener–Waterloo, Ontario, was driven by specialized manufacturing industries, not by scientific innovation from a research university. In time, the local business community caused the creation of local higher education institutions. It was only later that the University of Waterloo and its computer science and other research capacities evolved from the early business leadership of the industrial era. Portland, Oregon, is another case of a city-region that owes its high-tech economic development origins to business enterprise and not principally to the innovations of a local research university. Portland's high-technology economic development beginnings may be traced to innovations at the corporations, Tektronix and Intel. These are cases in "enterprise culture" (see this term, pp. 198–9).

Cities of Knowledge

Margaret Pugh O'Mara has used this label in her book of the same title (O'Mara 2005). She defined such cities as:

> These places were engines of scientific production, filled with high-tech industries, homes for scientific workers and their families, with research

> universities at their heart. They were the birthplaces of great technological innovations that have transformed the way we work and live, homes for entrepreneurship and, at times, astounding wealth. Cities of knowledge made the metropolitan areas in which they were located more economically successful during the twentieth century, and they promise to continue to do so in the twenty-first. . . . The city of knowledge was a creation of the Cold War, whose policies and spending priorities transformed universities, created vibrant new scientific industries, and turned the research scientist into a space-age celebrity. . . . The Cold War made scientists into elites, and mass suburbanization reorganized urban space in a way that created elite places. . . . The American research university was at the heart of this process, as economic development engine, urban planner, and political actor.
>
> (O'Mara 2005: 1–2)

O'Mara's book includes discussion of these cities of knowledge: Stanford University and the San Francisco Peninsula, The University of Pennsylvania and Philadelphia, and Georgia Tech and Atlanta. Compare these three cases of the Silicon Valley type of model to such modeling attempts by means of these six cases: Cambridge, England; Helsinki, Finland; Tel Aviv, Israel; Bangalore, India; Singapore; and Hsinchu-Taipei, Taiwan (Rosenberg 2002). In his book, *Cloning Silicon Valley*, David Rosenberg concluded that such cloning "is a difficult and delicate undertaking." Also, compare these cities of knowledge to the cities of business leadership and enterprise that were noted above in this section.

Clusters

Michael Porter has researched and written extensively on clusters and their significance in economic development:

> Clusters are geographic concentrations of interconnected companies and institutions in a particular field. Clusters encompass an array of linked industries and other entities important to competition. They include, for example, suppliers of specialized inputs such as components, machinery, and services, and providers of specialized infrastructure. Clusters also often extend downstream to channels and customers and laterally to manufacturers of complementary products and to companies in industries related by skills, technologies, or common inputs. Finally, many clusters include governmental and other institutions – such as universities, standards-setting agencies, think tanks, vocational training providers, and trade associations – that provide specialized training, education, information, research, and technical support.
>
> (Porter November–December 1998: 78)

Mike Brennan has developed a business-promotion working definition of "technology cluster" to identify and describe Michigan's technology clusters. The principal elements of this definition include: image internal to the cluster; perception of the cluster from outside the cluster; basic research is conducted in the cluster; a news mechanism that speaks for the cluster; a local-area organization that markets the cluster; a state-supported organization that promotes and facilitates technology cluster creation; and a publication that covers technology-based development and clusters statewide (Brennan July 2000: 1–2).

Within the context of the innovation systems approach, Davis and Schaefer use "'cluster' to refer to an aggregation of related firms [in a] geographically bounded area" (Davis and Schaefer 2003: 122). The rich empirical findings of David A. Wolfe and others teach us that clusters apply well in some contexts and less so in other contexts (Wolfe 2003; Bramwell *et al.* May 2004; cf. Ozawa 2004; Maskell and Kebir 2005).

Competitive Advantage Model

This model conceptualizes the various economic and social processes that confer more development success on some places and regions than others. A critical element of this model is the firm or business enterprise and its internal organizational behavior such as its motivation, strategy, management, commitment and competition. This model functions on the clustering and combining of the above processes at a location to affect productivity. In addition to these conditions, the approach of the model focuses on factor conditions, demand conditions, and related and supporting industries. From a strategic and planning perspective, the following factors of production can be created and it is the rate at which they are created that is critical: (1) human resources (quantity, skills, cost); (2) knowledge resources (technical, scientific and market knowledge); (3) capital resources (types, access, deployment), and (4) infrastructure (physical and social) [numbers added] (Plummer and Taylor 2001a: 227). Within the context of domestic and global demand, specialization may be a competitive advantage; this model complements and/or relates to several of the other local economic development theories noted in this section of the book, including flexible production, innovative milieus and learning regions. Plummer and Taylor have summarized Michael Porter's clustering of productivity in a place (cf. Porter 1990; November–December 1998) according to the competitive model as a function of social capital as influenced by local specialization and human resources; local demand and interregional trade; access to information; locational integration; institutional support; and technological leadership (Plummer and Taylor 2001a: 227). For related and comparative advantage US cases, refer to Annalee Saxenian's important book, *Regional Advantage* (Saxenian 1994).

Cosmopolite

In 1968, within the context of an early recognition of the shift from the city as a locally bound spatially circumscribed phenomenon to a node on the global network society, planner Melvin Webber introduced the notion of cosmopolite. He framed the concept in opposition to the notion of localite (see pp. 209–10). Cosmopolites are not tied to a specific place. Because of modern communication networks and extensive travel, their spatial range is global. Disconnected from location, they rely on information resources in their occupations and in their living. Cosmopolites "trade in information and ideas" (Webber Fall 1968: 1096).

Development

The ultimate impetus for regional and local planning in the global knowledge economy and network society is development – i.e. multifunctional, holistic, equitable and relational community development. Everett Rogers has offered an operational definition of such development that suits these purposes. While economic development, especially encompassing production and consumption functions, is the principal driving force, along with quality-of-life factors and community amenities, this operational definition provides the needed values imperative to address effectively the needs of economically distressed communities. Rogers defined development as:

> A widely participatory process of social change in a society intended to bring about both social and material advancement (including greater equality, freedom, and other valued qualities) for the majority of the people through their gaining greater control over their environment.
>
> (Rogers 1976: 225)

Digital Communities

These are networked places and spaces that are connected, developed, organized and have attained functional commonalities by means of information and communications technologies (ICTs). Such communities are composed of individuals, households, business firms and organizations. These elements converge functionally and cluster locationally or are spatially organized via networking to form digital communities. These communities range across a continuum of scale and places from neighborhoods and villages, to rural areas and countryside, to towns and cities, to metropolitan areas and networks of metropolitan areas, or a megalopolis, and to world-level and global-scaled connectivities that bind these elements and commonalities into ICT-integrated digital communities. They are primarily a local manifestation of the knowledge-based global economy.

Digital Development

This is the application of information and communications technologies (ICTs) infrastructure to community and economic development. The public policy goal of such digital infrastructure is directed "to fostering capital investment in a specific locale in order to create jobs and improve quality of life" (Laudeman May 16–19, 2002: 175). Compare this concept to Intelligent Development on pp. 205–6.

Digital Divide

This disparity is the "lack of access to IT for certain segments of the population" (Servon, 2002: 1).

Drama

From her experience in working with local community organizations and in doing advocacy planning in Cambridge, Massachusetts, and poor communities in South America, Lisa Peattie has conceived "community" and community organizing as a kind of drama or theatrical performance. She described a local planning project as follows:

> The community organizers were staging and directing; sometimes they edited a basically improvisational script. There were main actors, the officers and active members of the organization. And there was a supporting cast, those members of the community who could be gotten out to meetings and to public hearings and whose crowd noises at such occasions might intimidate the redevelopment authority. There were props, suggesting with some limited physical means a surrounding environment; the maps and reports our group was producing fell into this category.
>
> (Peattie November 1970: 406)

This analogue can be useful in enabling one to lay out the many relationships of the region's various actors and their roles as planning for local intervention proceeds for developing the future. See the section on Scenario, p. 216.

Economic Development

The US Economic Development Administration states that

> economic development is fundamentally about enhancing the factors of productive capacity – land, labor, capital, and technology – of a national,

> state or local economy. By using its resources and powers to reduce the risks and costs which could prohibit investment, the public sector often has been responsible for setting the stage for employment-generating investment by the private sector.
>
> (United States Economic Development Administration n.d.)

e-Envoy

The Cabinet Office states that:

> The primary focus of the Office of the e-Envoy is to improve the delivery of public services and achieve long term cost savings by joining-up online government services around the needs of customers. The e-Envoy is responsible for ensuring that all government services are available electronically with key services achieving high levels of use. The Office of the e-Envoy is part of the Cabinet Office of the Government of the United Kingdom.
>
> (Cabinet Office, Office of the e-Envoy n.d.)

Electronic Commerce

This retail activity also is popularly known as e-commerce. It broadly involves transactions over the Internet whether or not there is an exchange of money – e.g. e-government is a variant of e-commerce.

Electronic Readiness

"e-Readiness," operationally, has come to mean the extent to which an area's "business environment is conducive to Internet-based commercial opportunities. It is a concept that spans a wide range of factors, from the sophistication of the telecommunications infrastructure to the security of credit-card transactions and the literacy of the population" (cf. Economist Intelligence Unit 2005).

Enterprise Culture

A condition or pervasive attitude that exists in society, region, community, business or other organization in which a readiness to be entrepreneurial, innovative, to take initiative and to be self-reliant are valued and supported. This capacity, behavior and practice are shared widely among the stakeholders, institutional actors and individual citizens of the locality or the administrative structure (cf. Plummer and Taylor 2003). For examples of city-regions and their enterprise cultures, see the section on Cities of Business Leadership and Enterprise, p. 193.

Enterprise Segmentation and Unequal Power Relations

This model explains local growth as a function of three variables: (1) local control of technology; (2) the impact of large corporations; and (3) the network relationships of the locational integration of smaller firms. The nature of a location within power networks confers its enterprises with varying degrees of network centrality or peripherality. According to the model, greater network centrality may be attained by: (a) emphasizing export-led growth in contrast to a principal orientation to exploiting the local market; (b) providing skilled, stable jobs and continuous training; (c) early and rapid acquisition of new technologies combined with innovation and the retention of local inventions; and (d) a tendency toward geographical dispersion (cf. Plummer and Taylor 2001a: 228). These business-environment characteristics influence the potential of a locality to "generate and attract further enterprise, investment, and employment" (Plummer and Taylor 2001a: 228). It is the inequality and spatial unevenness among these characteristics across regions and localities that affect the business power relationships and therefore the competitiveness of location.

Espoused Theory

This concept is the theory of action that a person expresses as their own explanation and justification for behavior. This is in contrast to a person's "theory-in-use," which is the actual behavior. Simply, we usually do not do what we say we should do. The principal best practice of regional and local planning is both to have a well-planned espoused theory and theory-in-use. Refer also to Schön 1983; Argyris and Schön 1974.

Flexible Production and Flexible Specialization

The flexible production thesis explains regional economic development. It encompasses the issue of differential regional growth. Consequently, it is explicitly spatial. Plummer and Taylor further elaborate this model as consisting of regulation theory, institutionalist economics, evolutionary economics and transaction costs (Plummer and Taylor 2001a: 224). They have observed that "reduced transaction costs lie at the heart of these external economies" of tight-knit networks of the new industrial districts that are emerging as production areas of flexible production (Plummer and Taylor 2001a: 224).

> According to the theory, the new technological-institutional system of flexible production makes new locational demands with the result that new industries create their own spaces away from established centers of production and agglomerations of old industries, which then experience entropic death. Indeed, three forms of reagglomeration are said to have been developed in this most recent transition within capitalism: (1) craft-based, design-intensive centers (most notable the Third Italy); (2) high technology centers (for example, Silicon Valley); and (3) advanced producer and financial service agglomerations (for example, London). These industrial districts are, in turn, seen as being incorporated into a 'global mosaic of regional economies'.
>
> (Plummer and Taylor 2001a: 225)

In order to translate the technologically driven flexible production explanatory model into planned strategies for local economic development, these following forces need to be harnessed: (1) the local integration of firms by means of the exchange of goods and information, thereby reducing transaction costs; (2) close buyer-supplier relationships drive place-based technological leadership; (3) institutional capacities support; (4) a human resource base for the local labor market (Plummer and Taylor 2001a: 225). These are the ingredients that must be planned into local-area strategies when addressing the production segment of the e-business spectrum.

Framing

George Lakoff has defined frames as "mental structures that shape the way we see the world. As a result, they shape the goals we seek, the plans we make, the way we act, and what counts as a good or bad outcome of our actions" (Lakoff 2004: xv). Critical to the requirement of developing a new relational planning mindset is the need for regional and local planners and their client stakeholders to reframe the ends and means to a new future for the community that explicitly addresses the changed and changing conditions of the global knowledge economy that are impacting the locality.

Global Economy

"The global economy is based on the ability of the core activities – meaning money, capital markets, production systems, management systems, information – to work as a unit in real time on a planetary scale" (Castells May 9, 2001: 1). In varying combinations, the global economy and its forces of globalization both impact local regions and communities, and these forces represent development opportunities locally for strategic planning and intelligent development. Without effective local and regional planning and sustained commitment to plan implementation, the external impacts of globalization are likely to impact the local community in random or negative ways.

Good Governance

Fundamentally, good governance is about enhancing the functioning of the nature, quality and purpose of the totality of relationships that link various institutional spheres – local, state, civil society and the private sector – especially at the subnational – i.e. the city-regional scale. These relationships span formally structured and regulated dimensions and informal ones. Effective policy planning and policy implementation alignment – both vertically and horizontally – must be the goal of good governance. Effort invested in improving local and regional governance is quite likely to serve well the purposes of the planning practitioner community and its respective client groups. Refer to Kinuthia-Njenga n.d.

The United Nations Development Program (UNDP) has provided a list of behaviors and characteristics that can be used as a checklist to help frame the objectives and results of a good governance scenario and good planning. Planning scenarios for future states of good governance should have the following characteristics discussed and considered as criteria for action by a region's stakeholders: participatory; sustainable; legitimate and acceptable to the people; transparent; promotes equity and equality; promotes gender balance; tolerates and accepts diverse perspectives; able to mobilize resources for social purposes; strengthens indigenous mechanisms; operates by rule of law; efficient and effective in the use of resources; engenders and commands respect and trust; accountable; able to define and take ownership of solutions; enabling and facilitative; regulatory rather than controlling; able to deal with temporal issues; and service-oriented (German Foundation for International Development 2001).

Governance

Planning researcher Patsy Healey has provided us with a useful operational definition of governance, it is "... the modes of organization and practice through which collective affairs are managed; ... the affairs which are seen to be significant for cities, urban regions and territories (Healey July 2003b: 17). Governance is about the relationalities among the various levels of government, the private sector and civil society; and how those relationships affect and advance (or not) our communities.

> Governance involves interaction between the formal institutions and those in civil society. Governance refers to a process whereby elements in society wield power, authority and influence and enact policies and decisions concerning public life and social upliftment.
>
> (United Nations Economic and Social Commission for Asia and the Pacific 2002)

Governance is executed best by means of planned action and behavior taken collectively and individually.

Growth Poles and Growth Centers

This concept has evolved and developed to the stage where it may be characterized as consisting of innovating firms and industries that stimulate economic growth and its spread or spatial dispersion. These "propulsive industries" have been described as having the following developmental effects:

> 1 high degree of concentration;
> 2 high income elasticity of demand for their products, which are sold to a national [and global] market;
> 3 strong multiplier and polarization effects through input linkages;
> 4 an advanced level of technology and managerial expertise promoting local diffusion through demonstration effects;
> 5 promotion of a highly developed local infrastructure and service provision; and
> 6 the spread of 'growth-mindedness' and dynamism through the zone of influence.
>
> (Plummer and Taylor 2001a: 222)

Plummer and Taylor have extended the model to be dependent on three sets of processes: (a) large firms; (b) knowledge creation and transfer; and (c) new technology (Plummer and Taylor 2001a: 223). Some of the other theories of local economic development discussed here modify and drill deeper into these processes by portraying the importance of small and medium-sized firms, by emphasizing the spatial dynamics of innovative clusters and the Internet and geographical dispersion, and the strategic criticality of differentiating – by development stage – the characteristics of the industry life-cycle (Audretsch and Feldman 1996; Swann 1999). Refer to the section on Industry Life-cycle Model, pp. 203–4.

High Technology

Washington State's Technology Alliance has defined "technology-based business" as industries that

> have at least 10% of their employment in research and development (R&D) related occupations. This means that we define as "high-tech"

> sectors that commit significant resources to the development of new technologies, new products and new kinds of services. Interestingly, our definition has recently been used by the Bureau of Labor Statistics staff as a current definition of high tech.
>
> (Technology Alliance 2000)

Cortright and Mayer's study of the Portland, Oregon, metropolitan area's high technology industry used Standard Industrial Classification (SIC) codes for computer, electronics, instruments and software industries to define "high technology" (Cortright and Mayer, January 2001: 9). They also discuss the use of the newer classification system for industries – i.e. the North American Industry Classification System (NAICS).

Human Capital

Human capital is defined as a person's accumulation of knowledge, experience, skills, abilities, expertise and attitudes, the application of which may lead to the production of intellectual property, goods and services that may add value in the marketplace. Collections of such individuals may convey actual and potential power, growth and development to their respective units of association as in the cases of corporations and other organizations, households and regions and localities. Investments generate human capital by means of education, training, health care and development. The returns from human capital for the private individual take the forms of wages, salaries, other compensation, such as stock options and health benefits; human capital returns for society and the community include economic innovation, creativity and development. A major development goal for the region and locality is to invest in human capital sufficiently to generate creative and productive talent and to retain as much of that talent as possible.

Industry Life-cycle Model

The relationalities among the critical dimensions of available theory can be used to advance and steer, to some degree, the spillover effects of the research and intellectual property inherent in large investments into, for example, the life sciences and other sciences and technologies. These relations include time, space, and innovative science and technology content. Such relationalities may be planned by being informed by industry life-cycle theory. Audretsch and Feldman, in synthesizing the findings of others, have noted that:

> During the early stages of the industry life-cycle, there is a high amount of innovative activity and new and smaller enterprises tend to have the

> relative innovative advantage. During the mature stages of the industry life-cycle, there tends to be less (product) innovative activity, and established large enterprises tend to have the innovative activity advantage.
>
> (Audretsch and Feldman 1996: 256)

Thus, Audretsch and Feldman conclude that innovative firms entering in the early stages have the entrepreneurial technological advantage and these stages are relatively unfavorable to established firms. Alternatively, in the mature stages of the industry life-cycle, the incumbent firms generally have the routinized technological innovative advantage, with new entrants tending toward innovative disadvantage. Note that Amsden and Chu (2003) demonstrated how government policies, intervention and support were instrumental in upgrading mature high-tech industries and services. From this work, they derived new theory for use by policies planners seeking to apply the industry life-cycle model. See also the section on Product Life-cycle Model, pp. 213–14.

Information Age

Currently, we live and work in an era in which information increasingly is pervasive and easier to get. Indeed, information overload is a frequent complaint. Information has become so central and critical that it has become commonplace to perceive the economy and society of today respectively as an "information economy" and an "information society." The European Commission has been especially active in striving to ensure that Europe's citizens, businesses and governments "play a leading role in shaping and participating in the global information society" (Information Society Directorate-General 2002). The European Commission has formulated an eEurope action plan so as to achieve this goal.

Information and Communications Technologies (ICT)

The Organisation for Economic Co-operation and Development (OECD) has defined information and communications technologies (ICT) as industries that facilitate, by electronic means, the processing, transmission and display of information. Traditionally excluded are industries that create the information – i.e. the content industries. OECD's recent key ICT indicators are fifteen in number. Refer to Organisation for Economic Co-operation and Development 2005.

Innovation

For our purposes of advancing regional planning practice within the context of the global knowledge economy, innovation "is a locally driven process, succeeding where organization conditions foster the transformation of knowledge into products, processes, systems, and services" (Malecki 1997). In the context of economic development, economist Joseph Schumpeter (1934) early defined "innovation" as new business activities, especially along the principal domains of new products, new processes, new market, new materials and new organization. For an example of measuring sub-national innovation, see Policy One Research 2004.

Innovation Systems

For enabling an understanding of the dynamics involved in stimulating innovation for development, this is an important body of work. Innovation systems are the result of a complex set of relationships at various scales – e.g. national, regional and local such as the flow of information and the diffusion of information and communications technologies (ICTs) among networks of such actors as enterprises, institutions and individuals with the ultimate result being the generation of new, economically valuable knowledge – i.e. innovation. Such activities that occur within the borders of a nation state are called national innovation systems (NIS). Innovation systems at subnational scales – e.g. at the state or substate levels – have been termed regional innovation systems (RIS). At the sub-regional scale, those technology-related knowledge flows that produce or transfer new knowledge may be identified as local innovation systems (LIS). Compare to Local Innovation Systems (p. 209) and Regional Innovation Systems (p. 214).

Refer to Organisation for Economic Co-operation and Development 1997. This OECD document identified "four basic knowledge flows among actors in a national innovation system: (1) interactions among enterprises; (2) interactions among enterprises, universities and public research laboratories; (3) diffusion of knowledge and technology to firms; and (4) movement of personnel (Organisation for Economic Co-operation and Development, 1997: 12). Refer to the above section on Canada in Part III, pp. 75–8).

Intelligent Development

This is the ultimate form of development that should drive the practice of the knowledge-economy regional planner. Compare this concept to Digital Development (see p. 197). Intelligent development is digital development that explicitly draws on and is guided by theory, especially location theory and relational planning theory in the formulation of policies and the planning of communities for exploiting the potential of a place from the context of the global knowledge economy and network society. Intelligent development promotes investment in a region, resulting in wealth creation, human capital development, employment formation, creation of an enterprise culture and improvement in the quality of life. Such planning is focused on maximizing the value-added of a place by matching and segmenting the unique e-business functions, factors and relationalities of the region to the appropriate stages in the life-cycle processes of production and consumption functions being planned. Development is "intelligent," therefore, when the best practices from theory, benchmarking elsewhere and the appropriate applications of the latest technologies and best practices are utilized fully to develop a community holistically, multidimensionally and equitably. Simply, digital development is the means; intelligent development produces the ends. These two contemporary development forms are interdependent and self-reinforcing (cf. Foresight n.d.).

Knowledge

"Indigenous Knowledge," to paraphrase Jarboe, is local in that it is rooted in a particular region or place and situated within broader cultural traditions; it is a set of experiences generated by people living in these communities. It includes tacit knowledge and, therefore, is not easily codifiable. Tacit knowledge is transmitted orally, or through imitation and demonstration. Codifying it may lead to the loss of some of its properties. Tacit knowledge principally is experiential rather than theoretical knowledge – that is, it is earned through experience, and trial and error. Knowledge is learned through repetition, which is a defining characteristic of tradition even when new knowledge is added. Knowledge is constantly changing; it is being produced as well as reproduced, discovered as well as lost; although it is often perceived by external observers as being somewhat static (Jarboe May 9, 2002: 10).

Knowledge Economy

A knowledge economy is an economy that increasingly relies on technology and knowledge as factors of production and wealth creation, in addition to labor and capital. Technology and knowledge are transforming wealth-creation work from physically based to knowledge-based functions.

The knowledge base of an economy is "the capacity and capability to create and innovate new ideas, thoughts, processes and products and to translate these into economic value and wealth" (Huggins and Izushi 2002: 3). The Government of New Zealand's Ministry of Economic Development maintains an excellent Web page on "What Is the Knowledge Economy"; it includes a definition of the knowledge economy as "an economy which revolves around creating, sharing and using knowledge and information to create wealth and improve the quality of life" (Ministry of Economic Development n.d.).

Knowledge Society

The Government of New Zealand defines knowledge society as "a society where creating, sharing and using knowledge are key factors in the prosperity and well-being of people" (Ministry of Economic Development n.d.).

Knowledge Worker

Ahlen and Diggs (2003) define the knowledge worker as:

> The employee whose value to the employer is not embodied in skill, that is what the worker can do, but rather in what the worker knows, thinks, and communicates. In the new economy, the image of the knowledge worker coincides with information technology and its uses and influence on the firm. The knowledge worker, by the way, is not limited to a particular economic sector. There are knowledge workers in agriculture, manufacturing, administration, and virtually all sectors of the economy. In the new economy, knowledge-workers are critical for success.
>
> (Ahlen and Diggs 2003: 41–2)

Ls – the Five Ls

In order to put forward policy implications, David A. Wolfe, Co-Director of the University of Toronto's Centre for International Studies, has discussed the emerging findings of Canada's Innovation Systems Research Network (ISRN). He has classified them by the five Ls: learning, labor, location, leadership and legislation/labs (Wolfe January 22, 2004). Some of the development activities under each include:

> *Learning.* Human and intellectual capital development should occur in both old and new industries; education may occur both within and among the region's firms; both local firms and firms with corporate direction from outside the region must engage in continuous learning; learning for enhancing intellectual property development should be included in the region's learning strategy; and training designed for the production of specialized professionals is critical to the local human capital development program.
>
> *Labor.* Wolfe has stated that labor is the single most important input. He elaborated that "many places can produce 'talent,' but their ability to retain and attract talent depends on: thickness; opportunities; depth of local labor market; quality of place; creativity; diversity; and tolerance.

> *Location*. The research projects of the Innovation Systems Research Network found that concentrations of firms relied on strong linkages, both local and global.
>
> *Leadership*. Was found to be important at two levels, the firm and the community. Managerial talent and entrepreneurial ability differentiates one firm from another. For the community, "civic entrepreneurs" were found to play a role in creating clusters – e.g. Waterloo, see pp. 78–9 in Part III.
>
> *Legislation/labs*. The public sector, institutions, and regulations were found to play important roles in "shaping the rules of the game." Cited as important are "intellectual property rights, barriers to entry/exit, time horizons, labor market stability/mobility, immigration." The research of the Innovation Systems Research Network has demonstrated that universities and research institutions sometimes function to lead cluster formation, and other times they follow and support directions that have been taken by enterprises and business, e.g. Waterloo.
>
> (Wolfe January 22, 2004)

Learning Regions and Innovative Milieux

The "learning region" local economic development theory incorporates the place-based role of information, knowledge and learning – i.e. in the sense of tacit knowledge or the local exchange of knowledge that results in incremental innovation. "The availability of knowledge, especially tacit knowledge, is now seen as one of the strongest remaining spatially differentiated factors of production" (Plummer and Taylor 2001a). Learning regions and innovative milieux are concepts commonly used in Europe, Australia and Asia to describe local and regional economic systems that focus on knowledge and technology. The use of these terms is far less frequent in the US, where they are only occasionally used in academic settings. In a "learning economy," . . . "technical and organisational change have become increasingly endogenous. Learning processes have been institutionalized and feedback loops for knowledge accumulation have been built in so that the economy as a whole . . . is learning by doing" and "learning by using" (Lundvall and Johnson 1994: 26). In the "learning region," the term learning has a much broader meaning in that it refers to the collective and collaborative learning by all of the different actors in a region – i.e. learning from each other and learning with each other – in planning and implementing social and economic innovations (Stavrou 2003: 16).

According to Florida (1995):

> Regions are becoming focal points for knowledge creation and learning in the new age of global, knowledge-intensive capitalism, as they in effect become learning regions. The learning regions function as

> collectors and repositories of knowledge and ideas, and provide the underlying environment or infrastructure which facilitates the flow of knowledge, ideas and learning.
>
> (Florida 1995: 527)

Innovative Milieux are described by Camagni (1995) as:

> the set of relationships that occur within a given geographical area that bring unity to a production system, economic actors, and an industrial culture, that generate a localized dynamic process of collective learning and that act as an uncertainty-reducing mechanism in the innovation process.
>
> (Camagni 1995: 320)

Economic space becomes a "relational" space, in which actions through a synergetic and collective learning process reduce uncertainties during changes in technological paradigms. In addition, a collective learning process gives birth to creativity and continuous innovation (Camagni 1991: 1). In order to practice continuous learning in implementing the ALERT Model (see Part IV, pp. 131–3), the learning region and innovative miliex concepts should be adopted.

Local Innovation Systems

In order to understand better the distinctive characteristics of innovation at the local level, the Industrial Performance Center of the Massachusetts Institute of Technology initiated the Local Innovation Systems Project. The Center has defined local innovation systems:

> as spatial concentrations of firms (including specialized suppliers of equipment and services and customers) and associated non-market institutions (universities, research institutes, training institutions, standard-setting bodies, local trade associations, regulatory agencies, technology transfer agencies, business associations, relevant government agencies and departments, *et al.*) that combine to create new products and/or services in specific lines of business.
>
> (MIT Industrial Performance Center n.d.)

Compare to the section on Innovation Systems (p. 205) and Regional Innovation Systems (p. 214).

Localite

In contrast to the concept of cosmopolite (see p. 196), planner Melvin Webber wrote of the localite. He perceived these persons as individuals who live local lives. In an historical and/or ancestral sense, localities traditionally were rural, non-city based. Webber observed in 1968 that a reversal was underway. "Urbanites no longer reside exclusively in metropolitan settlements, nor do ruralites live exclusively in the hinterlands" (Webber Fall 1968: 1095). In the new global network society, Webber saw that similar to rural populations, those parts of the population that are least integrated into contemporary society are located in high-density areas in large city-regions. Isolated populations, whether rural or urban, are localites. They are tied to a specific place. Their spatial organization is local.

Mindset

Professional practice is rooted in teaching, experience and learning from the past. In this era of rapid change, the inertia from the past is a legacy on the planning professional's mindset and behavior that is likely to operate to impede new and flexible response and leadership, especially on the parts of practicing planners, economic developers and citizen planners. Refer to the section in Part II on The Challenge of Creating a New Mindset for Planning (pp. 42–3). For our purposes here, a University of Arizona course on the study of the future offers an applicable definition of mindset.

> A person's frame of reference that is fixed. A person can have a particular "mindset" that is so strong in a specific outlook that they do not see other perspectives, even though they might hear them and believe they have been given consideration. This prevents looking at new options in a realistic sense.
>
> (University of Arizona n.d.)

The *Changing*Minds.org website outlines the theories, principles, explanations, techniques, and among other topics, the disciplines involved in influencing mindset (*Changing*Minds.org n.d.). See the section on Path Dependence (pp. 211–12). Compare these notions with Donald Sull's concept of "active inertia," wherein leaders and managers of firms remain so committed to past approaches and strategies that have been successful for them, that they are unable to change with the times. Sull has developed operational actions that are intended to change status-quo behavior; these include strategic frames, relationships, processes, resources, and values (Sull 2003; July–August 1999).

Nanotechnology

This field

> uses devices created from individual atoms and molecules, and its advances have offered much promise. Demanding [on] all new manu-

> facturing processes, nanotechnology enables the fundamental building blocks of nature to be produced inexpensively – in virtually any arrangement. Almost any type of product is possible.
>
> (State Science and Technology Institute October 2002)

Network Society

A society where the key social structures and activities are organized around electronically processed information networks. These are "social networks which process and manage information and are using micro-electronic based technologies" (Castells May 9, 2001: 1).

Selecting quotations from Manuel Castells, he has written: "a network is a set of interconnected nodes. . . . Networks are open structures, able to expand without limits, integrating new nodes as long as they are able to communicate within the network . . ." (Castells 1996: 470). "The new economy is organized around global networks of capital, management and information, whose access to technological know-how is at the roots of productivity and competitiveness. . . . The network society, in its various institutional expressions, is, for the time being, a capitalist society. . . . But this brand of capitalism is profoundly different from its historical predecessors. It has two fundamental distinctive features: it is global, and it is structured to a large extent, around a network of financial flows" (Castells 1996: 471). "At a deeper level, the material foundations of society, space and time are being transformed, organized around the space of flows and timeless time" (Castells 1996: 476).

New Economy

This term

> describes aspects or sectors of an economy that are producing or intensely using innovative or new technologies. This relatively new concept applies particularly to industries where people depend more and more on computers, telecommunications and the Internet to produce, sell and distribute goods and services.
>
> (Government of Canada 2005)

Path Dependence

This concept illustrates "legacy planning" – i.e. the dominance of prior practice and the planner's general inability and sometime reluctance to change the current approach to doing planning because one has not been trained and or aware of the need to change practice behavior. For example, within the context of industrial clusters, path dependence introduces the importance of the past in technological and economic decision-making. In the development of a place, or a technology, or an industry, and the location of their beginnings can have a strong influence on their path of evolution and the likely pathways into the future. Simply, the current and future path of such development may be relatively dependent on the path of the past. See the section on Mindset (p. 210). For a case of multimedia and path dependence from Canada's Innovation Systems Research Network (ISRN) work, see Britton May 2004.

Planning Practice

Within the context of doing regional and local planning as part of the global knowledge economy and network society, "planning practices are an ensemble of social relations, networks and nodes of dynamic and often inventive social interaction, patterned by both legal, governmental and professional systems, and by customs and habits built up over the years" (Graham and Healey October 1999: 11). See the section on Practice, p. 213.

Planning Scenario

It is a set of planned activities that result in intentional future outcomes for a place. A planning scenario is a "willed" result, not principally a forecasted or probable future. The planning scenario goes beyond the descriptive or even the diagnostic, to the prescriptive. The planning scenario should offer directional suggestions and planned strategic-level pathways for the likely accomplishment of realistic, practical and desired future visions. To be realistic and "intelligent," planning scenarios must be driven, supported and reinforced by actual empirical evidence and guided by relevant concepts and theory. Refer to the section on Scenario, p. 216. For a recent case that used visioning and scenarios of community economic activity at the regional and local scale, refer to Marcus Grant, December 2004. For another useful scenario case at a continental scale and with a twenty-year time horizon that was developed in anticipation of the enlargement of the European Union, see Fink and Owen 2004. In his open systems planning design, the context of which is organization–consultant centered, G.K. Jayaram differentiates between three types of scenarios that should be executed in a sequence (Jayaram 1976). Conceptualized for the regional planning context, the open systems planning scenario design would be: (1) for the particular city-region, there is the creation of a *present scenario* by having the relevant regional actors identify, first the current external development-related expectations for the region and then the current internal development expectations of the region's stakeholders. The transactions across the external–internal boundary also are highlighted. At this point, value systems are incorporated explicitly into the open systems planning process discussions.

The planning group is asked the question, "Why does who expect what?" The answers begin from the perspective of each participant's personal values as reflected initially in the organizational context (Jayaram 1976: 279). (2) The creation of *realistic future scenarios* involves the representative actors and stakeholders of the region who are asked to perceive and project a specified future for the city-region, supposing that there was no deliberate intervention for change. (3) Regional actors then are asked to develop *idealistic future scenarios* for the city-region, describing the area for a specified future if interventions were made – i.e. what would the representatives like to see changed from the earlier scenarios that were identified. The idealistic future scenario is a good place to do "zero-base budgeting," for example (see p. 220). The idealistic future scenario construction also is the time for utopian thinking, as represented by earlier such traditions in planning – e.g. Goodman and Goodman 1947; Reiner 1963. Thereafter, the open systems planning actors further process all of the scenarios by identifying their perceived causes and relations, and temporally plan a select number of issues for the future in the region's strategic development. This temporal planning is intended to move the group to early action in the short-run by casting the issues in terms of acting tomorrow, six months from now and two years from now. This socio-technical open systems planning process illustrates the usefulness of discriminating between scenarios that characterize existing perceived realities in comparison to scenarios that are projected future realities and futures scenarios that are intended, desired and willed by the participants who represent some of the principal development interests of the region and its localities. The group processes for scenario development introduced here by Marcus Grant (refer to Scenario 7 pp. 162–5, in Part IV) and by G.K. Jayaram may be seen as rooted in the tradition of organizational and consultant-led change technologies that are facilitative of the principles of city-region relational planning and implementation (cf. Ogilvy 2002). These organizational-change technologies also are in the tradition of much of the work of Argyris and Schön (1974).

Practice

Practice is "a sequence of actions undertaken by a person to serve others, who are considered clients. . . . A theory of practice, then, consists of a set of interrelated theories of action that specify for the situations of the practice the actions that will, under the relevant assumptions, yield intended consequences" (Argyris and Schön, 1974: 6). See Planning Practice, p. 212).

Product Life-cycle Model

The central premise of the product life-cycle model is that, as technology matures through time, market conditions, production inputs, the nature and intensity of competition and location stability change. Aging of many technology products have demonstrated patterns of the relocation and outsourcing of the production of this technology from developed to developing countries. Product life-cycles are not inevitable; products can be repositioned for growth (Moon May 2005; Amsden and Chu 2003). Refer to the latecomer policies in the Taiwan case, discussed in Part III, pp. 87–8. See also the section on Industry Life-cycle Model, pp. 203–4.

Regional Innovation Systems

Philip Cooke has offered a definition of regional information systems "in which firms and other organizations such as research institutes, universities, innovation support agencies, chambers of commerce, banks, government departments, are systematically engaged in interactive learning through an institutional milieu characterized by embeddedness" (Cooke *et al.* 2004: 331). They are defined further as "interacting knowledge generation and exploitation sub-systems linked to global, national and other regional systems for commercializing new knowledge" (Cooke *et al.* 2004: 363). See Innovation Systems (p. 205) and Local Innovation Systems (p. 209). Refer also to the Innovative Systems section in Part III (pp. 76–8) and the work of David Wolfe and his research colleagues across Canada.

Relational Theory

Relational thinking and theorizing are congruent with the complex, multi-layered functions and flows of today's global knowledge economy and network society. This thinking can serve to stimulate our imagination and free up our collective and conventional neoclassical economic and locational perspectives of the world, and our approach to planning. For example, geographer Henry Wai-chung Yeung has discussed the "explanatory power in socio-spatial relations among such actors as individuals, firms, institutions, and other nonhuman actants" (Yeung March 13, 2002: 2). When relational theory is applied to, and practiced creatively within the context of, regional and local planning, it can be an effective means for advancing innovation and applying actionable knowledge to development, electronic business, and quality of life and community amenities. As used here, relational theory is the primary conceptual context for e-business planning and intelligent development. Relational theory is useful because it does not put sharp edges on concepts (Yeung March 13, 2002). Rather, it accommodates and explains the complicated relationships of the knowledge economy in realistic terms – i.e. terms that reflect well the empirical complexities of today's networked world.

In operationalizing relational theory for application to planning practice, planners Stephen Graham and Patsy Healey have argued for the translation of four interrelated points for effective contemporary planning practice:

> (1) consider relations and processes rather than objects and forms; (2) stress the multiple meanings of space and time; (3) represent places as multiple layers of relational assets and resources which generate a distinctive power geometry of places; and (4) planning practice should recognize how the relations within and between the layers of the power geometries of place are actively negotiated by the power of agency. [Numbers added]
>
> (Graham and Healey, 1999: 19–20)

For an elaborated discussion of relational theories within the context of contemporary urbanism, see Stephen Graham and Simon Marvin's book, *Splintering Urbanism* (2001). They identify relational theories as a "broad swathe of recent theoretical writing about cities and social change"; these writings "are diverse and notoriously difficult to define precisely" (Graham and Marvin 2001: 202). They identified two essential ideas from the relational theoretical literature on cities:

> relational urban and social theorists reject any idea that space, place and time have any essential, predefined or fixed meaning. . . . Time and space are thus both socially constructed together in all sorts of diverse ways within and through the contemporary metropolis, often via the uneven use of and connection to networked infrastructures that selectively help construct new social times and spaces.
>
> (Graham and Marvin 2001: 203)

and

> Cities and urban life, especially in today's heterogeneous, culturally mixed and polynuclear metropolitan areas, therefore need to be considered as 'multiplex' rather than 'uniplex' phenomena.
>
> (Graham and Marvin 2001: 204)

In order for these ideas to be incorporated fully and routinely into the practice of planning, changes in governance need to be embedded into the regulatory environments of localities and regions, as well as embedded in the daily behavior of planners and their principal stakeholders. It remains for these actors to apply the principles of relational planning creatively to the unique conditions and potentials of their respective communities. In order to practice relational regional planning, more than likely a changing and changed mindset is required. Refer to the "mindset" concept that was defined above.

Henry Yeung's more recent thinking on relational economic geography clarifies a highly abstract topic and makes it concrete and subject to being operational for planning practice and pragmatic action. He identified the essence of the core spatial and economic relationships: (1) actor-structure relationality; compare to the Program Planning Model (PPM) discussed in Part II, pp. 57–8;

(2) scalar relationality – i.e. global, national, regional and local; and (3) socio-spatial relationality – i.e. economic, social, political and spatial (Yeung 2005: 43). These relationalities are inherent in the e-business spectrum's functions and their linkages among those functions.

Research and Development

The US Department of Commerce Form RD-1A provides a useful definition of research and development (R&D); the US Census Bureau uses it as a collecting and compiling agent for the National Science Foundation of the United States: "R&D includes basic and applied research in the sciences and engineering. It also includes design and development of new products and processes and enhancement of existing products and processes" (US Department of Commerce 2002: 3).

Scenario

The scenario originated as an outline or synopsis of a play or other performance. It is used to develop a script that actors can follow to perform the play. Business, education (Vision 2010 n.d.) and other professions have evolved and adopted the technique as a means to produce a set of possible or hypothesized and sometimes fictional future events. Refer to the section on Drama (p. 197). The scenario can take various forms, from written outlines to narratives for storytelling and data-based analyses, to drawings, sketches, graphics (e.g. storyboards) and maps (i.e. of future probable intended conditions and spatial distributions). In recent years, information and communications technologies have been used to support the development of scenarios. For example, Fink and Owen have demonstrated the application of scenario software that assisted in sorting and ordering the many complexities and inconsistencies that occur among the influence factors identified by scenario conference participants in their case study (Fink and Owen Spring 2004). Also refer to the section on Planning Scenario (pp. 212–13) to contrast scenarios that generate possible and projected futures compared to planning scenarios that produce willed and intentional futures that are reinforced by strategic and tactical action plans (cf. Ogilvy 2002).

Science and Technology

Increasingly, science and technology are taking on greater importance in the economic growth of localities, regions, countries and globally. Science is the systematized method of knowing, discovering and deriving general laws or principles of behavior by means of experimentation, testing and replication about physical, natural and social phenomena. Technology is the practical application of knowledge. It provides a capability to apply knowledge, as in the case of computers facilitating calculations and information processing. For example, at the level of national large-scale projects, Japan has been investing in such science and technologies as nuclear energy, space development, aviation, marine development, life sciences, superconductivity, maglev trains, high-definition television, optical-fiber communications network and computer sciences (Ministry of Education, Culture, Sports, Science and Technology 2005). In the context of economic development planning, science and technology represent emerging opportunities for communities to grow and enhance the quality of life for its citizens. The strategic importance of science and technology is recognized at the national level. For example, the Executive Office of the President of the United States includes the Office of Science and Technology Policy (OSTP). The Director of OSTP co-chairs the President's Council of Advisors on Science and Technology, and oversees the President's National Science and Technology Council. Globally, for example, the Organisation for Economic Co-operation and Development (OECD) long has been concerned with determining and understanding the mechanisms of science and technology-driven policies in generating economic growth (Organisation for Economic Co-operation and Development 2001). The European Commission's Community on Research and Development Information Service (CORDIS) uses science and technology indicators to monitor progress on policies. Twenty-three such indicators are used on an annual basis. These range from research per 1,000 workforce, to total R&D expenditure in percent of GDP, to venture capital-investment per 1,000 GDP, to US patents per million population, and many more (Community Research and Development Information Service 2002).

Smart Zones

These are planned technology clusters. They are a state of Michigan government program

> intended to stimulate the growth of technology-based businesses and jobs by aiding in the creation of recognized technology-facilitated clusters of new and emerging businesses, those primarily focused on commercializing ideas, patents, and other opportunities surrounding corporate, university or private research institute R&D efforts. Smart Zones provide distinct geographical locations where technology-active firms, entrepreneurs and researchers can locate in close proximity to all of the community assets that will assist in their endeavors. The locations of the Michigan Smart Zones represent areas that comprise a critical mass of the technology development assets.
>
> (Michigan Economic Development Corporation n.d.)

There are 11 Smart Zones distributed across the state (see Figure 20, p. 121).

Social Capital

Social capital "refers to network ties of goodwill, mutual support, shared language, shared norms, social trust, and a sense of mutual obligation that people can derive value from" (Huysman and Wulf 2004: 1).

Stakeholders

The categories of institutional and individual interests in regions and localities include representatives of business: small and medium-size firms, large corporations, unions; government: various levels, both executive and elected; nonprofit civic organizations and institutions: educational, social welfare, healthcare, and so on; and individuals. Representative stakeholders are essential to operationalizing new governance behavior for the region. This changed behavior is a direct result of mindset change (see "mindset" above). Refer also to Part IV, the section on Stakeholders: Actors and Roles, pp. 104–7.

Talent

See the section on The Three Ts of Economic Growth below. Also refer to Human Capital on p. 203.

The Three Ts of Economic Growth

Richard Florida has written about the criticality of Technology, Talent and Tolerance working together and interdependently to produce creative places such as the San Francisco Bay Area; Boston; Washington, DC; Austin, and Seattle (Florida 2005a: 37–8). Florida operationally defined tolerance as openness, inclusiveness and diversity to all ethnicities, races and walks of life. Talent was defined as those with a bachelor's degree and above. Technology was defined as a function of both innovation and high technology concentrations in a region (Florida 2005a: 37). These activities and the importance of the relations among them have been revealed from the empirical research and resultant findings of Richard Florida and his colleagues.

Theory-in-Use

This is the theory of action that is the actual behavior of a person, in contrast to a person's "espoused theory," which is a person's expressed theory of action that a person uses to explain her or his actions and behavior. Simply, we usually do not do what we say we should do. The principal best practice of regional and local planning is both to have a well-planned espoused theory and a theory-in-use that are congruent with each other. Refer also to Schön 1983; Argyris and Schön 1974.

Translational Research

Being central to the core message of this book, translational research is driven by an explicit objective of advancing and learning from relational planning practice. Borrowing this concept from current clinical and basic medical research and practice, there is a need for more academic practitioner-planners to focus their scholarly effort on the intersection of theoretical and basic city-region planning research and empirical exploratory planning research and planning practice. Bi-directional feedback and exchange should be the rule at this interface. It is by means of reciprocal, reflexive interaction that the planning profession might more readily incorporate and internalize relational understanding and relational planning behavior, resulting in the realization of new relational mindsets and relational approaches by all planning practitioners. The Isle of Wight case discussed in Part IV (pp. 162–5) illustrates a benchmark regional and local planning project in translational research (Grant December 2004). Compare to the reflective practitioner concepts of Schön 1983.

Uniqueness of Place

In the context of taxonomic or classification systems, and at the theoretical level, "locations are not unique" (Bunge 1966: 375). However, in operational and pragmatic terms, in the study and in the planning of locations, each place and its region has its own set of distinctive and singular characteristics. It is the particular mix of attributes of a location that need to be taken into account in practicing planning and plan-implementation, especially in the context of the global knowledge economy and network society. Such recognition of a location's particularities enables development strategies to be formulated that are based on niche visioning and seeking to take advantage of the distinctive assets of a region and its localities. Simultaneously, planning practice today is enhanced by explicitly sharing development experiences and solutions to common problems. In this context, benchmarking and identifying best practices against which to assess the results of planned strategies should be central to monitoring and evaluating the success of the plan and the planning. Networking activities across places and regions can facilitate these activities in the development of a specific place. Compare this concept to the principles of the Competitive Advantage Model on p. 195 and to Brian Berry's divergent development pathways (Berry 1973).

Zero-base Budgeting

In the context of constructing planning scenarios for future intelligent development strategies there is value in freeing up one's thinking and in freshly approaching routine processes such as annual and short-term multi-year financial planning. One tactic that should be considered is to include zero-base budgeting planning in scenario planning exercises. "The zero-base budgeting (ZBB) method disregards the previous year's budget in setting a new budget, since circumstances may have changed. Each and every expense must be justified in this system" (Special Investor.com n.d.). Such a tactic can facilitate mindset change among planners, developers and regional stakeholders.

Gottmann Concepts, A–Z

Jean Gottmann was an early practitioner of city-region relational study and interpretation. He was explicit in identifying freshly and creatively relationalities of space, time, history, political geography and organizations, and resulting patterns of human behavior. For example, in elaborating on the white-collar revolution and the white-collar workers of the US northeastern megalopolis, he wrote, "their main work consists in relations with other people in the office or outside it, rather than in the operation of machines or the handling of materials" (Gottmann 1961: 627). He used metaphor effectively to convey the nuances and subtleties of these complex and new relational concepts. A selection of such relational concepts and metaphors is listed below.

Based initially on his mastery and perfection of the French geographic tradition (Buttimer 1971), Gottmann's innovative conceptual repertoire went well beyond this sampling below. His approach to geographic research resulted in spatial analyses based on historical facts and contemporary data interpretation. From his insightful interpretations of various relationships in and among places, he constructed useful and foresightful concepts. Even today, his pioneering approach offers us guidance on how to become aware of and to understand and conceptualize the conditions in which we live, work and play. His contributions are a foundation upon which we might build an intellectual strategy for achieving an improved, planned future.

Alexandrine Model

Reflecting on, and interpreting the teachings of Plato, Jean Gottmann derived the Platonic model of the Greek city. Plato conceived the "ideal polis" as one, the population of which should be located away from the sea (Gottmann 1984: 8). Additionally, Plato's ideal city was characterized as having "small scale, austerity, isolation, restricted maritime and trading activities [which will result in] a righteous and stable society" (Gottmann 1984: 9). Aristotle modified Plato's model by suggesting to his student, Alexander of Macedon, that instead of isolation, the unification of the Greeks could lead to their dominance in Asia as well as Europe (Gottmann 1984: 9). Alexander proceeded to conquer and build a vast connected empire that was antithetical to the Platonic model. The result was Greek dominance of much of Asia and Europe. These facts on the ground enabled Gottmann to derive and identify the Alexandrine model. This model was characterized as: "a vast, expanding pluralistic political and cultural system, bound together and lubricated by the active exchanges and linkages of a network of large trading cities" (Gottmann 1984: 10; Corey and Wilson July 2005a: 2).

Crossroads

Gottmann captured the essence of the shift of the city as the principal location for manufacturing and tertiary services to the place, a basic function of which:

> seems now replaced by one of servicing the ingathering of transients, most of whom have their residences and main places of employment elsewhere. In this way the city has now become essentially a vast crossroads that services chiefly people whom neither live nor are employed in it.
>
> (Gottmann November 1972: 308)

He observed that the increasing importance of these mobile populations:

> is due at least as much to transients from other parts within the country as to visitors (either for business or pleasure) from foreign lands. It is the "hub" function of the national or regional center that predominates in every substantial transactional district. The evolution of the structure of the labor force toward the quaternary activities is a major factor in the dynamics of large concentrations . . .
>
> (Gottmann November 1972: 308)

These growing quaternary occupations attract people who need face-to-face proximity and cultural amenities. This is a precursor to the tacit knowledge conceptualizations of today. These then-new crossroads functions were identified as including research institutions such as universities and large libraries; museums and other cultural institutions; trade centers; conference centers; large hotels with major conference facilities; growing employment in producer and business services, financial services such as banking, insurance, and real estate; and ritual activities such as mass recreation, religious pilgrimages, sports events and political conventions. Gottmann noted that the crossroads-city transient and mobile functions performed by the transactional city were not only for affluent transients. He observed that the underprivileged migrants and refugees were attracted to these new crossroads to take advantage of opportunities and welfare support; these transients also accounted for the new mobility (Gottmann November 1972: 309).

When Gottmann identified these megalopolitan and transactional dynamics, it was principally at the world-class and global-cities level. Since those early days, many of these functions have become important to the local economies of small and medium-sized city-regions around the world. He attributed many of these then-emergent new crossroads functions to special infrastructure, including technological evolution.

Evolution of Urban Centrality

Urban centrality "consists of a multiplicity of central functions gathered in one urban place; and it rests on one or several networks of means of access converging on that place" (Gottmann April 1975: 220). In pre-industrial periods the central functions consisted of administration in the form of the castle, commerce in the form of the market, and collective religious rituals in the form of the temple. Later, urban centralities followed cycles of production, first being concentrated in cities – i.e. fourteenth to seventeenth centuries – second, being scattered – i.e. late seventeenth to nineteenth centuries; third, concentrated again in the early twentieth century, and since the 1920s production is again dispersing while transactions are dominating center-city employment (Gottmann April 1975: 224; Corey 1980: 67).

External Change

External or exogenous change can render a place less important or more important. During the Classical period, for example, the island of Delos in the eastern Mediterranean Sea lost its centrality when the rise of Christianity functioned to marginalize Delos as the birthplace of Apollo. This external shift in religious favor had a dramatic shift in the internal fortunes of Delos. Today, for example, overseas outsourcing or other external forces of globalization can greatly change the local economy of a small manufacturing town in Western Europe or North America. Another example is when the national capital function is moved from one location to another; frequently in history, this has meant that the new destination, such as Brasilia, benefited, but it did not necessarily mean that Rio de Janeiro declined. Indeed, in the case of Brazil shifting its national capital, over time, both cities have grown and prospered.

Hinge

Jean Gottmann used the concept of hinge in diverse ways. Principally, in his urban interpretations, he used the hinge concept to illustrate the principal relations between place and network and the joining of these two realms. Among a number of other applications, he applied it as a locational hinge, an historical hinge and a pluralistic substantive hinge (Gottmann March/April 1983b: 89). He drew on both his history and geography training, in order to be positioned to integrate these two perspectives of time and space, and thereby becoming an extraordinary original conceptualizer of highly complex urban development and spatial-organizational power dynamics. For example, from Gottmann's observations of the history of Constantinople and Istanbul as a successful capital, he elaborated and further developed the concept of hinge. Indeed, he represented the relations of Constantinople as capital as a pluralistic hinge, in that it was a hinge between itself and the empire; Europe and Asia; the Romans and Greeks; land and sea power; present and past – i.e. an historical hinge; and different cultures and social structures, as in Christian and Moslem relationalities – i.e. a sociocultural hinge (Gottmann 1984: 11; Gottmann March/April 1983b: 89). Perhaps one of his most insightful applications of the hinge concept was in his landmark book *Megalopolis* (1961); it has a chapter entitled "The Continent's Economic Hinge." In this context, hinge was the metaphor for the link for the string of cities along the Northeastern Seaboard and the interior of the United States. These cities were at the interface of the continent and the network of trade relationships from the sea. This hinge-like system of seaport cities opened the trade door to the continent (Gottmann 1961: 103).

Hosting Environment

This is the set of amenities and the quality-of-life factors that are attractive to transactional actors – e.g. white-collar workers (Gottmann January/February 1979: 34–5). These factors enhance the centrality of the center city and operate to service the transactional transient and metropolitan resident alike (Corey 1980: 67).

Interweaving of Quaternary Activities

These are white-collar occupations and functions characterized by accessibility, information flows, transactional performance, quality of the labor market, amenities and entertainment, expert consultation, money and credit market, specialized shopping and educational facilities (Gottmann 1970: 329–30). Transactional webs result that operate within and between metropolitan areas, both nationally and internationally (Corey 1980: 67).

Megalopolis

The modern meaning of the word and concept is "a group of densely populated metropolitan areas that combine to form an urban complex. . . . According to Gottman[n], it resulted from changes in work and social habits" (The Columbia Encyclopedia 1993: 1738). His holistic regional interpretive skills enabled him to characterize the defining role and function of this US area of high population density and corporate organizational concentration as "the main street of the nation" (Gottmann 1961: 3; and cf. Leman Group 1976).

Orbits

The concept of orbits is one of the richest and more multifaceted of Gottmann's metaphors. Several of the concept's principal meanings and his applications connote movement through space as well as movement through time. Additionally, in a locational and spatial urban context, the orbit concept has multiple scales, spheres, reaches and different ranges. This meaning was a particular favorite of Gottmann's; he perceived the city both as "the center of a region, determined by local circumstances, or chiefly as a partner in a constellation of far-flung cities" (Gottmann 1984: 3). Further, he often used orbits in concert with the concept of "hinge." This enables one to see the city as the location, and joining of the local and regional urban orbit with the wider network connecting the local to distant multiple city locations and their orbits elsewhere. Gottmann characterized the local city-regional orbit as "secluded, insular;" he saw the widespread connected network of city-regions as "large" and "polynuclear" (Gottmann 1984: 4). He observed that one orbit – e.g. the Greek orbit – could be incorporated into another orbit – i.e. the Roman orbit. He observed that by means of "conquest, trade, settlement, road and city building," the Roman orbit was extended across three continents (Gottmann 1984: 10). This, and earlier "incorporations," such as Alexander's strategic empire building, enables one to construct a leadership and organizational orbit – i.e. an orbit that guides and gives vision, values, security, form and integrated coordination to other orbits. Especially from his observational and interpretational perspective of the Mediterranean region, Gottmann called attention to cultural diversities and conflicts, as in the case of the long-standing clash between the orbits of Islam and Christianity.

From his cross-regional experiences among the subglobal orbits of Europe, North America and Asia, Gottmann was able to compare and contrast the psychological and cultural orbits of proclivity and motivation. He raised hypotheses that the challenging physical environments of the Mediterranean region may have stimulated a degree of exploration and quest not practiced by other cultures, even though other cultures had similar knowledge and technology. "However, it is the Mediterranean-born culture that has swept around the planet and reorganized it in one orbit, diversified, partitioned, complicated as Mediterranean orbits always were, by now conscious of its unity." Thereby, he foretold the pervasiveness of globalization by stating, "the modern gradual evolution towards and interlocked world system is increasingly dependent on interwoven urban orbits" . . . that is, "structured by networks of transactions" (Gottmann 1984: 14).

Gottmann concluded his conceptually rich "orbits lecture" by stating:

> Some people like to travel more than others. The main lesson that Ulysses brought back from his wanderings was the knowledge of the diversity of the places he had visited and of their inhabitants. This is what he told Penelope about, when they were reunited. This is what the frescoes of prehistoric buildings at Akrotiri seem to describe. Does this ancient Mediterranean tradition express basic curiosity or even more, the impulse to learn how to deal with others, how to overcome distance and perhaps even how to overcome human diversity? The will of individuals to liberate themselves from their original environment even if it is the Garden of Eden? I am afraid I have got out of my prescribed orbit.
>
> (Gottmann 1984: 15)

From the many ways that Gottmann employed the concept and metaphor of orbits, he has left us with the enduring reminder that the relationalities of orbits constantly evolve and mutate. Therefore, currently successful regions, organizations or planning approaches cannot remain static; simply, he has taught us that we cannot afford to rest on our laurels. New orbits continue to emerge.

Platonic Model

See the Alexandrine model, pp. 221–2.

Quaternary Economic Activities

In his book *Megalopolis*, Gottmann discussed the growth of jobs in tertiary occupations and he elaborated on the white-collar subsector of the services sector. In that context, he introduced the concept of quaternary economic activities:

> Servicing modern production and consumption requires handling the goods on the one hand and the transactions on the other. Managing the transactions is no longer a simple matter of counting and contracting, or arithmetic and legal forms; today it involves information and research on the technology of products or of management and public relations, as well as on the events in the markets, local or distant information that cannot be obtained and used efficiently without education, competence,

> and special skills. Indeed, one wonders whether a new distinction should not be introduced in all the mass of nonproduction employment: a differentiation between tertiary services – transportation, trade in the simpler sense of direct sales, maintenance, and personal services – and a new and distinct quaternary family of economic activities – services that involve transactions, analysis, research, or decision making, and also education and government. Such quaternary types require more intellectual training and responsibility. The numerical increase in the tertiary and white-collar jobs appears to be related to an accompanying rise in the number of the professions and specializations classified under these older labels.
>
> (Gottmann 1961: 576–7)

Terms of Employment

These are the conditions that set "the time spent at work, during the day, the week, the year, the place of work, the nature of the services, the duration of the arrangement, the remuneration and benefits received by the employee" (Gottmann 1978: 393). Changes in the terms of employment permit more time and income for commuting, multiple places of residence, entertainment and continuing education, and so on, thereby contributing to the areal spread and the lowering of the density gradient of the transactional metropolis (Corey 1980: 67).

Transactional Forces

Gottmann conceptualized transactional forces. He demonstrated that employment is shifting from a labor force dominated by workers who produce and handle goods and tangible products – i.e. farmers, miners and factory craftsmen – to a labor force with a majority of its members engaged in the generation, processing and management of such intangibles as information, knowledge and decisions. The completion of these activities are transactions; the forces driving them are embodied in the transformation of economies and societies revolving around hardware to ones concerned increasingly with software. But the key question that Gottmann poses has to do with the role of cities in transactional economies and societies (Corey 1983: xi).

Planning Activities by Phase of the ALERT Model

Critical Planning Practice Activities of the *Awareness* Phase of the ALERT Model

The goal here is to develop reliable knowledge about the region's assets and resources.

The purpose of this goal is to study and generate specific knowledge of the region's economy and society, especially those conditions that relate to technology-facilitated and science- and technology-driven development – both current and potential.

The results of this goal should lay the groundwork for, and lead to actionable knowledge.

Increased awareness may be improved by study of the region and its localities – e.g. see Smart Michigan (n.d.) www.smartmichigan.org and access the sections on "Presentations" and "Reports" for examples of knowledge economy indicators being used to enable local stakeholders to become more aware of the structural changes underway in the regional and local economy.

Awareness, based on comparisons with other regions and localities – i.e. best practices and benchmarking, also can be particularly revealing and a basis for suggesting actions.

Lack of strategic awareness can be, and too often is widespread, even pervasive; planners, their governing commissions, and the public at large need to be more informed as to some of the structural changes that are impacting regions and localities in countries and regions around the globe.

Many tactics may be employed to achieve local awareness – e.g. the planners initially should develop factual profiles of the region with attention to the area's relationship to the global knowledge economy and network society; also citizen book clubs may be used to enhance awareness, both locally and globally – e.g. read Jared Diamond's book, *Collapse: How Societies Choose to Fail or Succeed* (2005).

In the end, with increased awareness, regional and local stakeholders can decide whether or not to act on their future; they may decide not to intervene, or they may decide to get smart, get activist and behave collaboratively in order to pursue an intelligent development path for the area's future.

Awareness therefore can lead to increased opportunities for the region – i.e. opportunities that otherwise would not be created without the investment of significant time and effort by the local stakeholders.

Critical Planning Practice Activities of the *Layers* Phase of the ALERT Model

The goal at this stage of applying the ALERT Model is to reconceptualize the practice of strategizing and planning for the future development of the region and its localities. Planners Graham and Healey have written that such reconceptualized planning practice should "represent places as multiple layers of relational assets and resources, which generate a distinctive power geometry of places," and this reconceptualization enables stakeholders and planners to "recognize how the relations within and between the layers of the power geometries of place are actively negotiated by the power of agency" (Graham and Healey October 1999: 20).

One of the most fundamental re-conceptualizations is to recognize that there are multiple meanings of space and time – e.g. there are many geographies to a place.

Maps and other graphics at various scales and times should be developed in order to reveal a systematic understanding of a place, and thereby to become aware of the area's many complex relations – i.e. relationships that are functional, spatial and temporal. See Smart Michigan (n.d.) www.smartmichigan.org and access the section on "Presentations," then click on "American Planning Association Intelligent Development Alert" for examples of knowledge economy mapped layers that may be used to enable local stakeholders to become aware of the diverse interrelationships among a region's e-business functions and their different geographies and histories; also see the graphics and maps in Corey and Wilson 2005b.

A way to operationalize the *layers* phase of the ALERT Model – i.e. functionally, spatially and temporally – has been to draw explicitly on relevant functional and locational economic development at the local and regional scales.

A result of such application was to enable the derivation of two principal factors that are imperative in the region's future spatial and strategic planning – i.e. (1) local human resources, and (2) local enterprise culture.

These two factors, and the many others that are deemed by the planners and stakeholders to be needed for full awareness, have spatial and time dimensions that should be understood as influencing the development of the place; in this context, these example elements have been used: for time and strategic planning, a relational planning program – e.g. the Program Planning Model, and the importance of the "moment" in timing and planning practice.

The key practical result of this phase of the ALERT Model is to have produced empirical facts of the region's economy in the form of its various critical multiple layers; thereby, the local planners and stakeholders of the region can become more aware of the area's linkages and distribution of knowledge economy resources, their development potential and thereby be empowered to formulate planned strategies and priorities for the future. Disparities among these various distributions can disclose the region's layers of power geometries; this can enable mediation and negotiation among these many power relations.

In the context of Michigan's planning regions, and in order to seek funding and grants from the US Economic Development Administration (EDA), stakeholders of the Economic Development Districts (EDD) and planning regions need to have formulated a Comprehensive Economic Development Strategy (CEDS). The CEDS is an example of "agency" for the realization of a planning instrument that enables the bringing together of the many layers and their power relationships into an instrument of strategy and action intended to produce "development" in the full sense of the word, that is to encompass a community's economic functions of production and consumption, as well as the essential cultural and social functions for a community's amenities and quality of life. In EDA's Investment Policy Guidelines, Assistant Secretary of Commerce for Economic Development, David Sampson, has established the following criteria – i.e. "The Sampson Seven" – for evaluating proposed EDA investments: (1) market-based; (2) proactive in nature and scope; (3) look beyond the immediate economic horizon, anticipate economic changes, and diversify the local and regional economy; (4) maximize the attraction of private sector investment that would not otherwise come to fruition absent EDA's investment; (5) have a high probability of success; (6) result in an environment where high skill, high wage jobs are created; and (7) maximize return on taxpayer investment (Sampson April 2, 2002). Regardless of country or region, this is an example of the kind of criteria-based investment guidelines that can be developed by local governance and used to prioritize, plan and monitor future digital development and intelligent development.

For context and operational purposes here, these functions and their relations have been included in an organizing framework that is termed the "Electronic Business Spectrum," or the "e-Business" Spectrum.

Critical Planning Practice Activities of the e-Business Phase of the ALERT Model

The goal of this phase of the ALERT Model is to operate from a working framework for organizing the general categories of knowledge economy functions and their complexities as they relate within the region and its localities and as they relate to the external world. This, the e-Business Spectrum framework, in this context, can be used both to study and analyze present economic conditions, as well as used for organizing the planning and intended priorities of the region's future development. These are the principal functions of the spectrum, including their respective dominant spatial organizational patterns; clustering or concentration patterns are indicated below by "C" in parenthesis, and dispersed or deconcentrated spatial patterns are noted by "D" in parenthesis.

Production functions

- Science- and technology-driven research and development (C);
- commercialization of products and services (C);
- business and producer services (C) and products (D);
- public and government producer services and products (C&D) – e.g. regulatory functions – fiscal and monetary decisions, information – e.g. weather forecasts, taxation, etc. via e-government.

Consumption (e-Commerce) functions

- online procurement: business to business and business to government (D);
- online retailing and e-government: business to consumer and government to consumer (D);
- value-added complementarities – e.g. electronic (clicks) sales and physical, or across-the-counter sales (bricks).

Amenity and quality of life factors

- social, cultural and institutional activities (C and D);
- natural environmental attributes (C).

Educational, talent and human capital capacity building, and high-quality for both (C and D)

See Part II, pp. 47–5 on Concepts, for a discussion of the elements of the e-business spectrum and the interrelationships among them.

Critical Planning Practice Activities of the Responsiveness Phase of the ALERT Model

The goal of this stage of the ALERT Model is to ensure that the planned change process as encapsulated by the Model is driven initially and periodically by the interpretations and priorities that may be derived and determined by the region's stakeholders with assistance from the local planners. These derivations may be made from the findings of the *awareness*, *layers* and *e-business* phases of the ALERT Model.

The focus of this phase is on the content and economic functions of the community; the focus may be framed, established and maintained by employing the e-business spectrum, its topics and the intended distributions, flows and linkages among the cells of the e-business spectrum.

The particular concern here is to move the community's development into the full range of the knowledge economy functions with explicit interest in ensuring that the area's digital infrastructure is widely distributed, thereby enabling the growth of intelligent development in the region and its localities. No area of the region should go unattended; each place has a role to play in the global knowledge economy and network society as represented by the e-business spectrum.

In order to move intelligent development visions and plans to actions and the realization of actual development opportunities, effective partnerships and complementary relationships must be forged. Here is where social capital investment can produce important returns in the context of ICT-enabled development (Huysman and Wulf 2004). Depending on the particular circumstances, both locally and regionally, such relationships need to span sectors, institutions and individuals. In short, new forms of governance must be devised, practiced and perfected. This will take time, gestation and maturation. It also will require strategies and plans for leadership, champions and divisions of labor. The results and timing of these intended actions require coordination and complementarity. As planned outcomes occur, however, they can serve increasingly as partial foundations from which to build and extend follow-on actions – i.e. the more pieces of the overall development strategy that are put in place, the clearer becomes the next set of actions that may be taken. In order to be responsive, attention must be paid to, and actions taken to address the unevenness of the distribution of digital infrastructure and e-business development opportunities that were noted in Part II in the discussion of "electronic business" (see pp. 47–51). Without such attention, inequities and economically distressed areas will become even more embedded in the new knowledge economy and network society of the regions of the state or province.

Responsiveness, to be most effective, needs to place primary emphasis on meeting the development demands of the community. The region's planners in particular need to listen, learn and incorporate their learnings into the strategic development planning for the community. There are several cases from North America that reveal different, but related lessons in responsiveness. In the Kitchener–Waterloo region of Ontario, Canada, for example, the local early twentieth-century entrepreneurs of the area's manufacturing sector were instrumental in ensuring the emergence of Waterloo Lutheran College. This evolved into the "Waterloo College Project," which led to the formation of the University

of Waterloo. This university's faculties ultimately initiated a world-class computer science program that has been determinant in moving the region's economic development into a position of global high-technology leadership. These outcomes were a direct result of a high level of civic spirit that evolved from common regional interest for promoting local growth (Bramwell *et al.* May 12–15, 2004: 6). The university-enabled Kitchener–Waterloo case may be compared and contrasted to the Portland and Oregon case. The Portland, Oregon–Vancouver, Washington metropolitan region, in contrast to widespread conventional wisdom, developed into a high-technology industry center without the benefit of a research university (Mayer and Provo, 2004: 16). The large firms of Tektronix and Intel, with their interests and capacities in research and development were critical in creating a regional innovation system despite the absence of a major research university. These firms and their network of suppliers and spin-offs have functioned as "surrogate universities" (Mayer February 17–20, 2002).

These two North American cases demonstrate that different roots and pathways may be used to produce responsive intelligent development results.

In order to ensure responsiveness to the many opportunities offered within the context of technology-facilitated and intelligent development, the region's governance needs to identify who is responding to such opportunities, and who remains to benefit by being even more responsive. This recognizes that actors in the community likely respond at different rates – e.g. businesses and younger members of the region may be early adopters of ICTs. Alternatively, other actors, such as local government, may need to be more responsive and proactive by having attractive and easy-to-use websites for economic development. The task here, then, is to identify who should be responding to what and plan a division of labor accordingly – e.g. the "who" may be firms, institutions and individuals such as those in the region with access and those with no access to high-speed broadband connectivity.

Critical Planning Practice Activities of the Talk Phase of the ALERT Model

The ultimate goal of this ongoing continuous and long-term phase of the ALERT Model is to plan and implement the digital development and intelligent development priorities that will have been led and proposed by the region's overall strategic planning steering group. *Talk*, in this context, stands for the diverse activities of the sustained regional and local planning and plan-implementation tasks required to enable the place to compete more effectively in the global knowledge economy and network society.

The transition from the *responsiveness* phase to the *talk* phase of the ALERT Model means getting beyond the initial frequent economic development promotion hype of technology-facilitated, and science and technology-based development programs, because now the hard work begins.

Identification, linkage to, and organization and incorporation of the region's instrumental e-business stakeholder groups are imperative. Collectively, they must practice innovation, collaboration, coordination, vision and leadership, thereby contributing to the development of a regional culture of creativity and entrepreneurship.

- The members of the stakeholder groups must be high-quality, credible, active – not passive citizens.
- Stakeholder groups will be transeconomic sector groups that are organized under the influence of the region and locality's umbrella strategic planning steering group.
- Stakeholders must develop a relational planning mindset.
- Stakeholder groups will recognize that mindset change requires time and socialization.
- Stakeholder groups will frame and construct action agendas and working divisions of labor.
- Stakeholder groups will conduct action research and commission studies, surveys and engage in training.
- In the process, there will be the development of an identity and sense of community for the region and its localities.
- Policy points are strategic outcomes of the ALERT Model process.
- In the longer term, policy points may evolve into new laws that express new collective values of the region.
- Stakeholders must develop a design for monitoring and evaluating new policies and policy outcomes of the ALERT Model, and issue a regular periodic scorecard of progress on the agendas of the community's stakeholders.
- Over time, the stakeholder groups will evolve and develop sets of shared common values on behalf of the region and its localities – e.g. reading books and plans together in order to deepen awareness of community assets and development potential, and to improve the quality of the public life being shared by the stakeholder groups.
- Stakeholder groups will evolve and develop public–private alliances and partnerships across the area's economic sectors from business to educational institutions to government to unions to selected local opinion leaders, and so on.

Outline of a Planning Scenario Approach to the Biosciences and the Program Planning Model in East Central Michigan Planning and Development Region (ECMPDR) by Karan Singh, 2004

Stage 1 – The Planning Mandate for Bioscience Industry-based Economic Development in the ECMPDR

- The planning mandate;
- the key stakeholders needed to implement the planning mandate and participate in regional economic development goals.

Stage 2 – The ECMPDR Problem Exploration

- Industrial classification and definition of the biosciences;
- firm level information and companies in Michigan in the ECMPDR involved in the biosciences;
- location of research institutions and four-year colleges/universities in Michigan, and their focus in the biosciences;
- external research monies to promote technology industries through institutional research;
- importance of venture capital resources, and Michigan's venture capital capacity;

- bioscience patents issued in Michigan's university towns between 1996 and 2002.

Stage 3 – Knowledge Exploration

- Best practices and benchmarking studies from states competing to attract bioscience industries;
- definition of the Product Life-cycle Model, towards understanding the Industry Life-cycle Model and the biosciences development cycle.

Stage 4 – The ECMPDR Program Design

- Prerequisites for growing a bioscience-based economy;
- the program design for growing a biosciences economy in the ECMPDR.

Stage 5 – The ECMPDR Program Activation

- Overview of the pilot program and implementation;
- pilot program implementation in the counties of Midland and Isabella within the ECMPDR using the scenario planning approach.

Stage 6 – The ECMPDR Program Operation and Diffusion

Stage 7 – The ECMPDR Program Evaluation

- Evaluation objectives for creating an attractive biosciences-based economy in the ECMPDR;
- evaluation criteria used to measure the success of the planned programmatic initiatives;
- program evaluation design;
- program evaluation analysis and feedback.

A Time-relational Method: The Program Planning Model

Relational theory encompasses both behavioral relations and processes over time. The Program Planning Model (PPM) is an empirically tested program planning and policy planning-process method that is compatible with relational theory (Van de Ven, and Koenig, Jr Spring–Summer 1976). One of its principal values is its effectiveness in breaking down complex policy and programmatic processes into understandable and actionable stages such that specialized actors may perform their stage-specific specialized roles. PPM facilitates a sequenced and iterative division of labor, assigned by phase, throughout a relational planned-change process through time.

The model offers planners and managers an effective means for introducing innovation and steering intentional planned strategic change in dynamic and uncertain societal and organizational environments, such as those environments that characterize the global knowledge economy and network society (Bryson 1988). PPM consists of seven principal stages, each of which has lead actors who perform specific roles; their relations complement the other roles and actors of the planning process.

Stage 1: *The initial mandate*. The role of this stage is to provide the official sanction of the organization or governance group of elites and representative regional stakeholders doing the planning to proceed with the initiation of this relational planning process. In Part IV, (pp. 97–185), the Regional Strategic Planning Steering Group discussed in the section on Stakeholders: Actors and Roles (see pp. 104–7), is a generic example of a new-governance entity that might convey a local mandate. The sanctioning should produce the mandate for the overall relational planning process. In the case of a formal statutory regional planning organization, the principal actors of this stage are the leaders of the organization(s), such as the board of directors and chief executive officer, etc. (Delbecq and Van de Ven March 1971). Early in the process, a planning coordinating committee or steering group should be formed that includes representatives of the region's principal stakeholder groups that are intended to play roles throughout the planning process, including especially representatives of those most directly impacted, such as consumers, clients, citizens and the implementers of the resulting plans, programs and services. A start-up resource allocation enables the planning work to begin.

Stage 2: *Problem exploration*. The role of this stage is to identify the region's problems that are to be addressed by the planned-change process. These problems should be in the form of measurable needs – i.e. the facts and data must be identified that will be used to drive the following stages of the planning process. Surveys, focus groups and other group techniques may be used to identify the locality's particular problems and needs – e.g. Delbecq *et al*. 1986. The principal actors of this phase are the consumers or clients with the needs, and their role here is to provide an articulation, elaboration and specification of the problems as they see them. Data analyses and surveys are used by the planners to explore the problems. The planning coordinating and steering committee reviews the outcomes of this stage and sanctions (or not) progression to the next planning phase.

Stage 3: *Knowledge exploration*. The role of this stage is to provide external information on state-of-the-art best practice solutions to the empirical problems and needs that were generated in the previous stage. The primary actors at this stage are experts, especially experts external to the planning organization or steering group. The contribution of the external experts is to offer informed ideas that can be translated into planned actions to address the identified consumer or client needs. The key function here is to infuse the planning process with contemporary ideas and problem solutions that are fresh to the thinking and practices of the local strategists. The planning coordinating committee or steering group reviews the outcomes of this stage and approves movement of the planning effort to the next phase (or not).

Stage 4: *Proposal development*. The function of this stage is to produce an initial strategic proposal outline that can serve as the preliminary conceptual framework for an integration of the needs and knowledge from Stages 2 and 3. This integration must be configured so that it provides the program planners with the material for developing thorough, strategic and comprehensive programmatic solutions. The principal actors here are local program planners; other contributing actors are internal and external resource controllers and the organization's or group's formal leader-elites from Stage 1. The planning coordinating committee reviews the outcomes of this stage and approves (or not) progression of the planning effort to the next stage. A further allocation of budgeted resources may occur at the end of this stage.

Stage 5: *Program design*. The outcome of this stage is a program-level plan – i.e. one that guides focused action at operational and tactical levels. Again, the key actors are the program planners; they conduct the technical program design tasks. The planning coordinating committee reviews the outcomes of this stage and endorses (or not) the implementation of the plan in the form of three substages of plan-execution: pilot, demonstration and full implementation. Plans for staged implementation enables experimentation and resulting feedback; it also avoids making big mistakes by limiting the scope of plans before moving to larger, higher-impact plans. Enhanced funding for such implementation is critical at this stage.

Stage 6: Program implementation and program transfer. The role of this stage is to begin the execution of the stage's program-implementation design. The most effective means of implementing a new program or innovation is to behave in systematic, experimental and formative ways. This includes, initially, the execution of a small-scale pilot project and the incorporation of the lessons from this activity into a larger-scale demonstration test project and, depending on these results and their assessment, the learnings from the demonstration may be integrated into a full-scale implementation of the planned program design. The principal actors here again are the program planners, along with representatives of program implementers and service providers of the planned program. Representative stakeholders continue to provide oversight and contribute to the planning process. The planning coordinating committee and steering group reviews the substage lessons and outcomes of this stage.

Stage 7: Program evaluation. Program evaluation activities parallel Stages 4–6. The role of this activity is to produce and execute a monitoring and evaluation plan that provides regularly scheduled routine audits to ensure that the problems and needs from Stage 2 are being effectively addressed and reduced respectively. These activities enable the organization or group and its clients or consumers to know to what degree the program plan is being successful or not; if not, then mid-course corrections or termination of the program can take place on an informed and measurable basis. The primary actors here are the program planners, outside evaluators and representatives of the region's consumers, citizens or clients; their role is to contribute to these assessments, review them and advise on the implications of the findings.

In sum, the PPM method is designed to introduce innovation – i.e. new planned programs. The model is useful especially because it has been tested thoroughly in many different planning practice settings, and organizational and institutional environments. This process approach and its embedded relationships have been found to produce effective results. Simply put, this relational planning-process model works (Corey 1988).

References

Note: A number of the references do not display the date of publication. The references with "no date" are marked as (n.d.). Suggestions to "compare" issues and cases are abbreviated by the use of (cf.).

Ackroyd, P. (2001) *London: The Biography*, London: Vintage.

Ahlen, J.W. and Diggs, M. (2003) *The Keys to Growth in the New Economy: Investing in Discovery, Engineering, and Entrepreneurship*, Little Rock: Arkansas Science and Technology Research Authority. Available at www.arkansasscienceandtechnology.org/pdfs/keys_to_growth.pdf (accessed July 8, 2005).

Albrechts, L., Healey, P. and Kunzmann, K.R. (2003) "Strategic Spatial Planning and Regional Governance in Europe," *Journal of the American Planning Association*, 69 (2): 113–29.

Americans for the Arts (2003) *Arts and Economic Prosperity: The Economic Impact of Nonprofit Arts Organizations and their Audiences*, Washington, DC: Americans for the Arts.

Amsden, A.H. and Chu, W.W. (2003) *Beyond Late Development: Taiwan's Upgrading Policies*, Cambridge, MA and London: The MIT Press.

Andersen, B. and Corley, M. (2003) "The Theoretical, Conceptual and Empirical Impact of the Service Economy: A Critical Review," United Nations University Discussion Paper No. 2003/22. Available at www.unu.edu/hq/library/Collection/PDF_files/WIDER/WIDERdp2003.22.pdf (accessed August 23, 2003).

Anderson Economic Group (AEG) (2004) "The Life Sciences Industry in Michigan: Employment, Economic and Fiscal Contributions to the State's Economy," report presented to Michigan's Core Technology Alliance, Lansing, Michigan: 1–27.

Appalachian Regional Commission (n.d.) "Appalachian Regional Commission: Welcome." Available at www.arc.gov (accessed July 4, 2005).

Argyris, C. and Schön, D.A. (1974) *Theory in Practice: Increasing Professional Effectiveness*, San Francisco, CA: Jossey-Bass.

Association of Southeast Asian Nations (n.d.) "Toward an e-ASEAN." Available at www.aseansec.org/7817.htm (accessed May 25, 2005).

Atkinson, R.D. (1998) *The New Economy Index: Understanding America's Economic Transformation*, Washington, DC: The Progressive Policy Institute, November. Available at www.neweconomyindex.org (accessed September 6, 2000).

—— (1999) *The State New Economy Index*, Washington, DC: Progressive Policy Institute, July. Available at www.neweconomyindex.org (accessed December 6, 2001).

—— (2002) *The 2002 State New Economy Index: Benchmarking Economic Transformation in the States*, Washington, DC: Progressive Policy Institute, June. Available at www.neweconomyindex.org/states/2002/index.html (accessed September 27, 2004).

—— (2005) *The Past and Future of America's Economy: Long Waves of Innovation that Power Cycles of Growth*, Northampton, MA: Edward Elgar.

Atkinson, R.D. and Gottlieb, P.D. (2001) *The Metropolitan New Economy Index: Benchmarking Economic Transformation in the Nation's Metropolitan Areas*, Washington, DC: The Progressive Policy Institute. Available at www.neweconomyindex.org (accessed April 25, 2001).

Audretsch, D.B. and Feldman, M.A. (1996) "Innovative Clusters and the Industry Life Cycle," *Review of Industrial Organization*, 11: 253–73.

Aurigi, A. (2005) "New Technologies, Yet Same Dilemmas? Policy and Design Issues for the Augmented City." Paper presented at the Digital Communities 2005 Conference, Benevento, Italy, June.

Baum, H.S. (1986) "Politics in Planners' Practice," in B. Checkoway (ed.) *Strategic Perspectives on Planning Practice*, Lexington, KY: Lexington Books, pp. 25–42.

—— (1996) "Practicing Planning Theory in a Political World," in S. Mandelbaum, L. Mazza and R.W. Burchell (eds) *Explorations in Planning Theory*, New Brunswick, NJ: Center for Urban Policy Research, Rutgers, The State University of New Jersey, pp. 365–82.

—— (2000) "Communities, Organizations, Politics, and Ethics," in C.J. Hoch, L.C. Dalton and F.S. So (eds) *The Practice of Local Government Planning*, 3rd edn, Washington, DC: International City/County Management Association, pp. 439–64.

Baumol, W. and Bowen, W. (1966) *The Performing Arts: The Economic Dilemma*, New York: Twentieth Century Fund.

Beito, D.T., Gordon, P. and Tabarrok, A. (eds) (2002) *The Voluntary City: Choice Community, and Civil Society*, Ann Arbor, MI: The Independent Institute and The University of Michigan Press.

Bell, D. (1974) *The Coming of Post Industrial Society*, London: Heineman.

Bennis, W. (1989) *Why Leaders Can't Lead: The Unconscious Conspiracy Continues*, San Francisco, CA: Jossey-Bass.

Berry, B.J.L. (1973) *The Human Consequences of Urbanisation: Divergent Paths in the Urban Experience of the Twentieth Century*, New York: St Martin's Press.

Biegel, S. (2001) *Beyond our Control? Confronting the Limits of Our Legal System in the Age of Cyberspace*, Cambridge, MA: MIT Press.

Booz Allen Hamilton (2002) *The World's Most Effective Policies for the e-Economy: International e-Economy Benchmarking*, London: Booz Allen Hamilton, November 19. Available at www.itis.gov.se/publikationer/eng/ukreport.pdf/ (accessed May 23, 2005).

Bramwell, A., Nelles, J. and Wolfe, D.A. (2004) "Knowledge, Innovation and Regional Culture in Waterloo's ICT Cluster," paper presented at the ISRN National Meeting, Simon Fraser at Harbourfront Centre, Vancouver, BC, May 12–15.

Braverman, H. (1974) *Labor and Monopoly Capital: The Degradation of Work in the Twentieth Century*, New York: Monthly Review Press.

Brennan, M. (2000) "Technology Clusters and Emerging Technology in Michigan: A Report for CyberState." Available at www.cyber-state.org/4_0/techcluster.pdf (accessed July 7, 2005).

Brenner, N. (2004) *New State Spaces: Urban Governance and the Rescaling of Statehood*, New York: New York University.

Breuckman, J.C. (2003) *An Examination of Government-Led Initiatives in Michigan*, Community and Economic Development Occasional Papers, Lansing, Michigan: Community and Economic Development Program, Michigan State University, April. Available at www.cedp.msu.edu/PDF%20FILES/BrueckmanAppendices.pdf/ (accessed May 23, 2005).

Britton, J.N.H. (2004) "The Path Dependence of Multimedia: Explaining Toronto's Cluster," paper presented at the Sixth Annual Innovative Systems Research Network (ISRN) National Meeting, Vancouver, BC, May.

Brody, S.D. (2003) "Measuring the Effects of Stakeholder Participation on the Quality of Local Plans Based on the Principles of Collaborative Ecosystem Management," *Journal of Planning Education and Research*, 22 (4): 407–19, Spring.

Bromley, R. (2002) "Doxiadis and the Ideal Dynapolis: The Limitations of Planned Axial Urban Development," *Ekistics*, 69 (415/416/417): 317–30, July/August, September/October, November/Decemer.
—— (2003) "Towards Global Human Settlements: Constantinos Doxiadis as Entrepreneur, Coalition-Builder and Visionary," in J. Nasr and M. Volait (eds) *Urbanism Imported or Exported*? Chichester and New York: John Wiley & Sons, pp. 316–40.
Brookings Metropolitan Policy Program (n.d.) "Weekly Update." Available at www.brookings.edu/metro (accessed July 5, 2005).
Brooks, M.P. (2002) *Planning Theory for Practitioners*, Chicago and Washington, DC: Planners Press, American Planning Association.
Bryson, J.M. (1988) *Strategic Planning for Public and Nonprofit Organizations: A Guide to Strengthening and Sustaining Organizational Achievement*, San Francisco, CA: Jossey-Bass.
Buck, N., Gordon, I., Hall, P., Harloe, M. and Kleinman, M. (2002) *Working Capital: Life and Labour in Contemporary London*, London and New York: Routledge.
Bunge, W. (1966) "Locations are Not Unique," *Annals of the Association of American Geographers*, 56: 375–6.
Bunnell, T. (2004) *Malaysia, Modernity and the Multimedia Super Corridor: A Critical Geography of Intelligent Landscapes*, London and New York: RoutledgeCurzon.
Bunnell, T., Barter, P.A. and Morshidi, S. (2002) "Kuala Lumpur Metropolitan Area: A Globalizing City-Region," *Cities*, 19 (5): 357–70.
Buttimer, A. (1971) *Society and Milieu in the French Geographic Tradition*, Washington, DC: Association of American Geographers.
Cabinet Office (n.d.) "Office of the e-Envoy." Available at www.e-envoy.gov.uk (accessed July 7, 2005).
Cairncross, F. (1995) "The Death of Distance," *The Economist*, 336 (7934), September 30: S1-S28.
—— (1997) *The Death of Distance*, Boston, MA: Harvard Business School Press.
Camagni, R. (1991) "Introduction: from the Local 'Milieu' to Innovation Through Cooperation Networks," in R. Camagni (ed.) *Innovation Networks: Spatial Perspectives*, London: Belhaven.
—— (1995) "The Concept of Innovative Milieu and its Relevance for Public Policies in European Lagging Regions," *Papers in Regional Science*, 74 (4): 317–40.
Canada's Technology Triangle (n.d.) "Canada's Research Triangle." Available at www.tech-triangle.com (accessed July 10, 2005).
Castells, M. (1989) *The Informational City: Information Technology, Economic Restructuring and the Urban-Regional Process*, Oxford: Blackwell.
—— (1996) *The Rise of the Network Society*, Cambridge, MA: Blackwell.
—— (2001) "Identity and Change in the Network Society: Conversation with Manuel Castells," in *Conversations with History* Series, Berkeley, CA: Institute of International Studies, University of California at Berkeley: 1–4, May 9. Available at globetrotter.berkeley.edu/people/Castells/castells-con4.html (accessed January 7, 2003).
Center for Arts and Cultural Policy Studies, Princeton University, NJ (n.d.). Available at www.princeton.edu/~artspol (accessed November 24, 2003).
*Changing*Minds.org. (n.d.) "Welcome." Available at changingminds.org (accessed September 21, 2005).
Choe, S.C. (1996) "The Evolving Urban System in North-East Asia," in F.C. Lo and Y.M. Yeung (eds) *Emerging World Cities in Pacific Asia*, Tokyo: United Nations University Press, pp. 498–519.
—— (1998) "Urban Corridors in Pacific Asia," in F.C. Lo and Y.M. Yeung (eds) *Globalization and the World of Large Cities*, Tokyo: United Nations University Press, pp. 155–75.
Clark, C. (1940) *The Conditions of Economic Progress*, London: Macmillan.
Clark, P.A. (1972) *Action Research and Organizational Change*, London: Harper & Row.

Clarke, S.E. and Gaile, G.L. (1998) *The Work of Cities*, Minneapolis, MN and London: University of Minnesota.
Cohen, R., Schaffer, W. and Davidson, B. (2003) "Arts and Economic Prosperity: The Economic Impacts of Nonprofit Arts Organizations and their Audiences," *Journal of Arts Management, Law and Society*, 33: 17–31.
Cohill, A. and Kavanaugh, A. (1997) *Community Networks: Lessons from Blacksburg, Virginia*, Norwood, MA: Artech House.
Colgan, C.S. (2004) "Measuring and Tracking New Regional Economics: The 'Creative Economy' and the 'Ocean Economy'," paper presented at the Association of Collegiate Schools of Planning, Portland, OR, October.
The Columbia Encyclopedia (1993) *The Columbia Encyclopedia*, 5th edn, New York: Columbia University Press.
Community Research and Development Information Service (2002) "Science and Technology Indicators for the European Research Area." Available at www.cordis.lu/rtd2002/indicators/ind_at.htm (accessed July 8, 2005).
Connect Northwest (n.d.) "Connect Northwest." Available at www.connectnw.org (accessed July 4, 2005).
Cooke, P., Heidenreich, M. and Braczyk, H.J. (eds) (2004) *Regional Innovation Systems: The Role of Governance in a Globalized World* (2nd edn), London and New York: Routledge.
Corey, K.E. (1980) "Transactional Forces and the Metropolis: Towards a Planning Strategy for Seoul in the year 2000," in *The Year 2000: Urban Growth & Perspectives for Seoul*, Seoul: Korea Planners Association, pp. 54–89.
—— (1983) "An Introduction to the Transactional City," in J. Gottmann, *The Coming of the Transactional City*, College Park, Maryland, MD: University of Maryland Institute for Urban Studies, pp. xi-xvii.
—— (1987) "Planning the Information-Age Metropolis: The Case of Singapore," in L. Guelke and R. Preston (eds) *Abstract Thoughts: Concrete Solutions*, Essays in Honour of Peter Nash, Waterloo, Ontario: Department of Geography, University of Waterloo, pp. 49–72.
—— (1988) "The Program Planning Model: A Tool for Policy Planning and Policy Research in South Korea and Sri Lanka," in F.J. Costa, A.K. Dutt, L.J.C. Ma and A.G. Noble (eds) *Asian Urbanization Problems and Processes*, Berlin and Stuttgart: Gebruder Borntraeger, pp. 86–92.
—— (1991) "The Role of Information Technology in the Planning and Development of Singapore," in S.D. Brunn and T.R. Leinbach (eds) *Collapsing Space and Time: Geographic Aspects of Communication and Information*, London: HarperCollins Academic, 1991, pp. 217–31.
—— (1998) "Information Technology and Telecommunications Policies in Southeast Asian Development: Cases in Vision and Leadership," in V.R. Savage, L. Kong and W. Neville (eds) *The Naga Awakens: Growth and Change in Southeast Asia*, Singapore: Times Academic Press, pp. 145–201.
—— (2000) "Intelligent Corridors: Outcomes of Electronic Space Policies," *Journal of Urban Technology*, 7 (2): 1–22.
—— (2002) "Electronic Commerce and Digital Opportunity for Local, Urban and Regional Development Planning," Community and Economic Development Occasional Paper, Lansing, MI: Center for Urban Affairs, Michigan State University. Available at www.ssc.msu.edu/~espace/SmartMich.htm (accessed May 30, 2005).
—— (2003) "Planning for Life-Culture: New Planning Theory for Planning Practitioners," in *Life-Culture and Regional Development Planning: Making Gyeonggi a Livable Place*, Suwon, Korea: World Life-Culture Forum, Gyeonggi, pp. 69–108, December 19.
—— (2004a) "Moving People, Goods, and Information in Singapore: Intelligent Corridors," in R.E. Hanley (ed.) *Moving People, Goods, and Information in the Twenty-First*

Century: *The Cutting-Edge Infrastructures of Networked Cities*, London and New York: Routledge, pp. 293–324.
—— (2004b) “Relocation of National Capitals: Implication for Korea,” in *International Symposium on the Capital Relocation*, Seoul: Seoul Development Institute, pp. 42–127.
—— (2004c) “Capital Relocation in Korea: A Case for Choice and Options in Planning Practice (Part II),” paper prepared for the International Symposium on the Capital Relocation, Seoul, Seoul Development Institute, September 22.
Corey, K.E. and Wilson, M.I. (2005a)”European Influence in Shaping Urban and Regional Planning: A Dream for the Globe,” paper presented at the Association of European Schools of Planning, Vienna, Austria, July.
—— (2005b) “The Naga Matures: From IT to Intelligent Development Policies in Southeast Asia,” in V.R. Savage and M. Tan-Mullins (eds) *The Naga Challenged*: *Southeast Asia in the Winds of Change*, Singapore: Marshall Cavendish Academic, pp. 301–46.
—— (2005c) “Improving Urban and Regional Technology Planning in the Global Knowledge Economy and Network Society: Toward a Global Planning Practice Learning Community,” paper presented at the Eighth International Asian Planning Schools Association Conference, Penang, Malaysia, September. Available at www.apsa2005.net (accessed September 27, 2005).
—— (2006 forthcoming) “Toward Good e-Governance for the Future in Southeast Asia,” in M. Rao and I. Banerjee (eds) *New Media and Development in the Asia-Pacific*: *Regional Perspectives*, Singapore: Marshall Cavendish Academic.
Cortright, J. and Mayer H. (2001) “High Tech Specialization: A Comparison of High Technology Centers,” Survey Series, Washington, DC: Center on Urban and Metropolitan Policy, The Brookings Institution, January.
—— (2002) *Signs of Life*: *The Growth of Biotechnology Centers in the U.S.*, Washington, DC: Center on Urban and Regional Policy, The Brookings Institution, June.
Countryside Agency (n.d.) “The Countryside Agency.” Available at www.countryside.gov.uk (accessed June 27, 2005).
CPANDA (Cultural Policy and the Arts National Data Archive), Princeton University. Available atwww.cpanda.org (accessed November 24, 2003).
Craig, J. and Greenhill, B. (April 2005) *Beyond Digital Divides? The Future for ICT in Rural Areas*, London: Commission for Rural Communities.
Daniels, P.W. (1985) *Service Industries*, London: Methuen.
Davis, C.H. and Schaefer, N.V. (2003) “Development Dynamics of a Start-Up Innovation Cluster: The ICT Sector in New Brunswick,” in D. Wolfe (ed.) *Clusters Old and New*: *The Transition to a Knowledge Economy in Canada’s Regions*, Montreal and Kingston: McGill-Queens University Press, pp. 121–60.
Delbecq, A.L. and Van de Ven, A.H. (1971) “The Generic Character of Program Management: A Theoretical Perspective,” Madison, WI: Center for the Study of Program Administration, University of Wisconsin, March.
Delbecq, A.L. Van de Ven, A.H. and Gustafson, D.H. (1986) *Group Techniques for Program Planning*: *A Guide to Nominal Group and Delphi Processes*, Middleton, WI: Green Briar Press.
DeRuyter, R. (2004) “Growing the BlackBerry: Research In Motion’s Mike Lazaridis Explains Company’s Strategy for Reaching New Markets,” *The Record*: B1 and B3.
Development Gateway (n.d.) “Governance.” Available at http://topics.developmentgateway.org/governance (accessed June 28, 2005).
Diamond, J. (2005) *Collapse*: *How Societies Choose to Fail or Succeed*, New York: Viking.
Dionne Jr, E.J. (2005) “Visions of the New New Orleans,” *washingtonpost.com*, September 20. Available at www.washingtonpost.com/wp-dyn/content/article/2005/09/19/AR2005091901296.html (accessed September 26, 2005).
Dodge, M. and Kitchin, R.M. (2000) *Mapping Cyberspace*, London: Routledge.
—— (2001) *Atlas of Cyberspace*, Reading, MA: Addison-Wesley.

Dowall, D.E. and Whittington, J. (2003) *Making Room for the Future: Rebuilding California's Infrastructure*, San Francisco, CA: Public Policy Institute of California.
Doxiadis, C.A. (1966) *Between Dystopia and Utopia*, Hartford, CT: The Trinity College Press.
—— (1967) *Emergence and Growth of an Urban Region: The Developing Urban Detroit Area*, Vol. 2: Future Alternatives, Detroit, MI: The Detroit Edison Company.
—— (1968) *Ekistics: An Introduction to the Science of Human Settlements*, London: Hutchinson.
—— (1969) *Emergence and Growth of an Urban Region: The Developing Urban Detroit Area*, Vol. 1: Analysis, Detroit, MI: The Detroit Edison Company.
—— (1970) *Emergence and Growth of an Urban Region: The Developing Urban Detroit Area*, Vol. 3: A Concept for Future Development, Detroit, MI: The Detroit Edison Company.
—— (1977) *Ecology and Ekistics*, Boulder, Colorado: Westview Press.
Doxiadis, C.A. and Papaioannou, J.G. (1974) *Ecumenopolis: The Inevitable City of the Future*, New York: W.W. Norton.
Drennan, M. (2002) *The Information Economy and American Cities*, Baltimore, MD: Johns Hopkins University Press.
Droege, P. (ed.) (1997) *Intelligent Environments: Spatial Aspects of the Information Revolution*, Amsterdam: Elsevier.
Drucker, J. and Li, Y. (2005) "Phone Giants are Lobbying Hard to Block Towns' Wireless Plans," *The Wall Street Journal*: A, June 23.
Economist Intelligence Unit (2005) "The 2005 e-Readiness Rankings: A White Paper from the Economist Intelligence Unit." Available at graphics.eiu.com/files/ad_pdfs/2005 Ereadiness_Ranking_WP.pdf (accessed July 5, 2005).
Ellis, S., Hirmis, A. and Spilsbury, M. (2002) *How London Works*, London: Kogan Page.
Errington, S., Stocklmayer, S.M. and Honeyman, B. (2001) *Using Museums to Popularise Science and Technology*, London: Commonwealth Secretariat.
European Commission (1994) Europe and the Global Information Society – Recommendations to the European Council, Bangemann Report, Brussels: European Commission.
Eurostat (2003) *Education Across Europe 2003*, Luxembourg: Office for Official Publications of the European Communities. Available at www.epp.eurostat.cec.eu.int/cache/ITY_OFFPUB/KS-58-04-869/EN/KS-58-04-869-EN.PDF (accessed August 24, 2005).
—— (2004) *Work and Health in the EU: A Statistical Portrait*, Luxembourg: Office for Official Publications of the European Communities.
Evans, P. and Wurster, T.S. (2000) *Blown to Bits: How the New Economics of Information Transforms Strategy*, Boston, MA: Harvard Business School Press.
Fahey, L. and Randall, R.M. (eds) (1998) *Learning from the Future: Competitive Foresight Scenarios*, New York and Chichester: John Wiley & Sons.
Faludi, A. and Waterhout, B. (2002) *The Making of the European Spatial Development Perspective: No Master Plan*, London and New York: Routledge.
Feketekuty, G. (1987) *International Trade in Services*, Cambridge: Ballinger.
Fink, A. and Owen, M. (2004) "Scenarios for the Future of Europe's Regions," *Futures Research Quarterly*, 20 (1) (Spring): 5–27.
Fisher, A.G.B. (1939) "Production, Primary, Secondary and Tertiary," *The Economic Record*, June: 24–38.
Fletcher, R.G., Moscove, B.J. and Corey, K.E. (2001) "Electronic Commerce: Planning for Successful Urban and Regional Development," in J.F. Williams and R.J. Stimson (eds) *International Urban Planning Settings: Lessons of Success*, International Review of Comparative Public Policy, Vol. 12, Amsterdam: JAI and Elsevier Science, pp. 431–67.
Florida, R. (1995) "Towards the Learning Region," *Futures*, 27 (5): 527–36.
—— (2002a) "The Economic Geography of Talent," *Annals of the Association of American Geographers*, 92 (4): 743–55.

—— (2002b) *The Rise of the Creative Class: And How It's Transforming Work, Leisure, Community and Everyday Life*, New York: Basic Books.
—— (2004) "America's Looming Creativity Crisis," *Harvard Business Review*, 82 (10): 122–36.
—— (2005a) *Cities and the Creative Class*, New York and London: Routledge.
—— (2005b) *The Flight of the Creative Class: The New Global Competition for Talent*, New York: HarperBusiness.
Florida, R. and Gates, G. (2001) "Technology and Tolerance: The Importance of Diversity to High-Tech Growth," Washington, DC: Center on Urban and Metropolitan Policy, The Brookings Institution.
Flynn, P. and Hodgkinson, V.A. (eds) (2001) *Measuring the Impact of the Nonprofit Sector*, New York: Kluwer Academic/Plenum Publishers.
Foresight. (n.d.) "Welcome to Intelligent Infrastructure Systems." Available at www.foresight.gov.uk/Intelligent_Infrastructure_Systems/ (accessed July 6, 2005).
Frederick, E. (2005) "Michigan: A Recipe for e-Government Development," *Bulletin of the American Society of Information Science and Technology*, 31 (20): 1–10, December/January. Available at www.asis.org/Bulletin/Dec-04/frederick.html (accessed May 16, 2005).
Friedman, T.L. (2005a) *The World Is Flat: A Brief History of the Twenty-First Century*, New York: Farrar, Strauss and Giroux.
—— (2005b) "Observing Katrina from Singapore," *The Straits Times*: 21, September 15.
Gandhi, J., *et al.* (2005) "Metropolitan Michigan Knowledge Economy Indicators," Lansing, MI: Michigan State University Community and Economic Development Program, June.
Garreau, J. (1991) *Edge City: Life on the New Frontier*, New York: Doubleday.
GaWC (Globalization and World Cities Study Group) (n.d.) "Globalization and World Cities Study Group and Network." Available at www.lboro.ac.uk/gawc (accessed July 4, 2005).
German Foundation for International Development (DSE) (2001) "Good Governance and Health," Issue Paper in Preparation for the International Colloquium Cotonou/Benin, Bonn, Germany: German Foundation for International Development.
Gershuny, J. (1978) *After Industrial Society: The Emerging Self Service Economy*, London: Macmillan.
Gladwell, M. (2000) *The Tipping Point: How Little Things Can Make a Big Difference*, Boston, MA, New York and London: Little, Brown & Company.
—— (2005) *Blink: The Power of Thinking Without Thinking*, New York and Boston, MA: Little, Brown & Company.
Goodman, P. and Goodman, P. (1947) *Communitas: Means of Livelihood and Ways of Life*, Chicago: The University of Chicago Press.
Gottlieb, P.D. (1994) "Amenities as an Economic Development Tool: Is There Enough Evidence?" *Economic Development Quarterly*, 8 (3): 270–85.
—— (1995) "Residential Amenities, Firm Location and Economic Development," *Urban Studies*, 32 (9): 1413–36.
Gottmann, J. (1961) *Megalopolis: The Urbanized Northeastern Seaboard of the United States*, New York: The Twentieth Century Fund.
—— (1970) "Urban Centrality and the Interweaving of Quaternary Activities," *Ekistics*, 29 (174): 322–31.
—— (1972) "The City is a Crossroads," *Ekistics*, (204): 308–9, November.
—— (1974) "The Dynamics of Large Cities." *Geographical Journal*, 140 (2): 254–61.
—— (1975) "The Evolution of Urban Centrality," *Ekistics*, 39 (233): 222–8, April.
—— (1978) "Urbanisation and Employment: Towards a General Theory," *Town Planning Review*, 49 (3): 393–401.
—— (1979) "The Recent Evolution of Oxford," *Ekistics*, 46 (274): 33–6 January/February.
—— (1983a) *The Coming of the Transactional City*, College Park, MD: University of Maryland Institute for Urban Studies.
—— (1983b) "Capital Cities," *Ekistics*, 50 (299): 88–93, March/April.

—— (1984) *ORBITS: The Ancient Mediterranean Tradition of Urban Networks*, The Twelfth J.L. Myres Memorial Lecture, lecture delivered at New College, Oxford, May 3, London: Leopard's Head Press.
Goulet, D. (1983) *Mexico: Development Strategies for the Future*, Notre Dame and London: University of Notre Dame Press.
Government of Canada (2005) "Economic Concepts New Economy." Available at canadianeconomy.gc.ca/english/economy/neweconomy.html (accessed July 8, 2005).
Graham, S. (2002) "Flow City: Networked Mobilities and the Contemporary Metropolis," *Journal of Urban Technology*, 9 (1): 1–20.
—— (ed.) (2004) *The Cybercities Reader*, London and New York: Routledge.
Graham, S. and Healey, P. (1999) "Relational Concepts of Space and Place: Issues for Planning Theory and Practice," *European Planning Studies*, 7 (5): 623–46, October.
Graham, S. and Marvin, S. (1996) *Telecommunications and the City: Electronic Spaces, Urban Places*, London and New York: Routledge.
—— (2001) *Splintering Urbanism: Networked Infrastructures, Technological Mobilities and the Urban Condition*, London and New York: Routledge.
Grant, M. (2004) "Innovation in Tourism Planning Processes: Action Learning to Support a Coalition of Stakeholders," *Tourism and Hospitality: Planning and Development*, 1 (3): 219–37, December.
Greenstein, R. and Wiewel, W. (eds) (2000) *Urban–Suburban Interdependencies*, Cambridge, MA: Lincoln Institute of Land Policy.
Hall, P. (1997) "Megacities, World Cities and Global Cities," the First Megacities Lecture, Rotterdam, the Netherlands. Available at www.megacities.nl/lecture_1/lecture.html (accessed July 4, 2005).
Hall, P., *et al.* (2003) "POLYNET: Sustainable Management of European Polycentric Mega-City Regions, An Application to INTERREG IIIB NWE Secretariat ENO, Community Initiative," Lille, France: European Regional Development Fund.
Hamnett, C. (2003) *Unequal City: London in the Global Arena*, London and New York: Routledge.
Hampton, K.N. (2003) "Grieving for a Lost Network: Collective Action in a Wired Suburb," *The Information Society*, 19: 417–28.
Hanak, E. and Baldassare, M. (eds) (2005) *California 2025: Taking on the Future*, San Francisco, CA: Public Policy Institute of California.
Harrison, B. and Bluestone, B. (1988) *The Great U-Turn: Corporate Restructuring and the Polarizing of America*, New York: Basic Books.
Harvey, D. (1989) *The Condition of Postmodernity*, Oxford: Basil Blackwell.
Healey, P. (2001) "Polycentric Development." Available at www.esprid.org/expert.asp?expertid=1 (accessed August 16, 2004).
—— (2002) "Collaborative Planning: Shaping Places in Fragmented Societies," in G. Bridge and S. Watson (eds) *The Blackwell City Reader*, Malden, MA: Blackwell, pp. 490–501.
—— (2003a) "Planning in Relational Space and Time: Responding to New Urban Realities," in G. Bridge and S. Watson (eds) *A Companion to the City*, Malden, MA: Blackwell, pp. 517–30.
—— (2003b) "Network Complexity and the Imaginative Power of Strategic Spatial Planning" paper draft prepared for presentation for the ACSP/AESOP (Association of Collegiate Schools of Planning/ Association of European Schools of Planning) Congress, Leuven, Belgium, July.
—— (2003c) "European Context of Technical and Economic Co-operation Between Northern Ireland and the Republic of Ireland," presentation made at the Dublin Belfast Corridor 2025 Conference, Newry, Northern Ireland, September 11.
—— (2004a) "The Treatment of Space and Place in the New Strategic Spatial Planning in Europe," *International Journal of Urban and Regional Research*, 28 (1): 45–67.
—— (2004b) "The Role of Collaborative Practices in Governance Transformations: Experiences in City Region Spatial Planning in Europe," paper prepared for presentation for

the Association of Collegiate Schools of Planning (ACSP) Conference, Portland, OR, October 20–4.
—— (2005) "Urban Region Strategy-Making in a Relational World," paper presented at the Association of European Schools of Planning (AESOP) Congress, Vienna, July.
—— (2006 forthcoming) "Toward Good E-governance for the future in Southeast Asia," in M. Rao and I. Banerjee (eds) *Media and Development in the Asia-Pacific*, Singapore: Marshall Cavendish Academic.
Hepworth, M. (1990) *Geography of the Information Economy*, New York: Guilford Press.
Ho, K.C., Kluver, R. and Yang, K.C.C. (2003) *Asia.com*: *Asia Encounters the Internet*, London and New York: RoutledgeCurzon.
Hodge, G. and Robinson, L.J. (2001) *Planning Canadian Regions*, Vancouver and Toronto: UBC Press.
Holbrook, J.A. and Wolfe, D.A. (eds) (2000) *Innovation, Institutions and Territory*: *Regional Innovation Systems in Canada*, Montreal and Kingston: McGill-Queens University Press.
—— (eds) (2002) *Knowledge, Clusters and Regional Innovation*: *Economic Development in Canada*, Montreal and Kingston: McGill-Queens University Press.
Holzheimer, T. (2005) "Benchmarking the Creative Class within the Washington, DC Metropolitan Area," *News & Views American Planning Association Economic Development Division*, 1: 8–12, Winter.
Horowitz, M. (2002) "Growing Your Own: Building Blocks for BIO-Based Economies," paper presented at the pre-conference workshop of the 6th Annual Conference of the State Science and Technology Institute, Dearborn, Michigan, October 1.
Huggins, R. and Izushi H. (2002) *World Knowledge Competitiveness Index 2002*: *Benchmarking the Globe's High Performing Regions*, Cardiff, Wales: Robert Huggins Business & Economic Policy Press.
Huggins, R. *et al.* (2004) *World Knowledge Competitiveness Index 2004*, Pontypridd, Wales: Robert Huggins Associates.
Huysman, M. and Wulf, V. (eds) (2004) *Social Capital and Information Technology*, Cambridge, MA: The MIT Press.
Infocomm Development Authority (2005) "IDA Unveils its 10-year Infocomm Technology Roadmap," *Singapore Wave*, 15: 4–5.
Information Society Directorate-General of the European Commission (2002) "Mission Statement." Available at europa.eu.int/comm/dgs/information_society/mission/index_en.htm (accessed July 8, 2005).
Institute of Portland Metropolitan Studies (n.d.) *Silicon Forest Family Tree and Silicon Forest Universe*: *A Visual History of the High Tech Industry in Metropolitan Portland-Vancouver* (a poster), Portland: Institute of Portland Metropolitan Studies, Portland State University, Available at www.upa.pdx.edu/IMS/currentprojects/siliconforest.html (accessed June 19, 2005).
Intellect (n.d.) "Technology – Enabling the UK Knowledge Driven Economy." Available at www.intellectuk.org/camaigns/knowledge/intellect_index.asp (accessed June 21, 2005).
International Labour Organization (2004) *World Employment Report 2004–05*. Available at www.ilo.org/public/english/employment/strat/wer2004.htm (accessed August 24, 2005).
International Telecommunication Union (2003) "ITU Digital Access Index: World's First Global ICT Ranking." Available at www.itu.int/newsroom/press_releases/2003/30.html (accessed December 14, 2003).
—— (2005) *Basic Indicators*. Available at www.itu.int/ITU-D/ict/statistics/at_glance/basic03.pdf (accessed August 24, 2005).
I-Ways (2005) "Implementing the e-APEC Strategy," *I-Ways, Digest of Electronic Commerce Policy and Regulation*, 28: 45–57.
Jacobs, J. (2005) *Dark Age Ahead*, New York: Vintage Books.
Jarboe, K. (2002) "Using the Tools of the Information Age," presentation at the Economic Development Administration Chicago Region 2002 Economic Development Conference, May 9.

Jayaram, G.K. (1976) "Open Systems Planning," in W.G. Bennis, K.D. Benne, R. Chin and K.E. Corey (eds) *The Planning of Change*, 3rd edn, New York: Holt, Rinehart & Winston, pp. 275–83.

Kellerman, A. (1993) *Telecommunications and Geography*, London: Belhaven Press.

—— (2002) *The Internet on Earth: A Geography of Information*, Chichester: John Wiley & Sons.

Kim, A.J. (1993) "A Model of Unified Spatial Development in the Korean Peninsula," paper prepared for the International Conference on Transformation in the Korean Peninsula Toward the 21st Century: Peace, Unity and Progress, East Lansing, MI, Michigan State University, July 7–11.

King, Jr, M.L. (1965) "Remaining Awake through a Great Revolution," commencement address for Oberlin College, Oberlin, OH. Available at www.oberlin.edu/external/EOG/BlackHistoryMonth/MLK/CommAddress.html (accessed January 5, 2005).

Kinuthia-Njenga, C. (n.d.) "Good Governance: Common Definitions." Available at www.unhabitat.org/HD/hdv5n4/intro2.htm/ (accessed May 17, 2004). Search on "good governance common definitions."

Komninos, N. (2002) *Intelligent Cities: Innovation, Knowledge Systems and Digital Spaces*, London and New York: Spon Press.

Kunzmann, K.R. (2004) "Unconditional Surrender: The Gradual Demise of European Diversity in Planning," paper presented at the Association of European Schools of Planning Congress, Grenoble, France. Available at www.planum.net/topics/main/m-kunzmann-epp.htm (accessed January 18, 2005).

Lakoff, G. (2004) *Don't Think of an Elephant! Know Your Values and Frame the Debate*, White River Junction, VT: Chelsea Green Publishing.

LaMore, R.L., Gandhi, J., Melcher, J., Supanish-Goldner, F. and Wilkes, K. (2004) "Michigan Knowledge Economy Index: A County-Level Assessment of Michigan's Knowledge Economy," Lansing, MI: Michigan State University Community and Economic Development Program, July. Available at www.smartmichigan.org (accessed June 21, 2005). Click on "Reports".

—— (2005) "Snapshots of Port Huron's Knowledge Economy," presentation at the Citizens Planner Workshop for St. Clair County, April. Available at www.smartmichigan.org (accessed June 21, 2005). Click on Reports.

Landry, C. (2000) *The Creative City: A Toolkit for Urban Innovators*, London: Earthscan Publications.

Lang, R.E. (2003) *Edgeless Cities: Exploring the Elusive Metropolis*, Washington, DC: Brookings Institution Press.

Laudeman, G. (2002) "TechSmart: A Catalytic Approach to Digital Development," in R. Carveth, S.B. Kretchmer and D. Schuler (eds.) *Shaping the Network Society: Patterns for Participation, Action and Change*, Seattle: Proceedings of the Directions and Implications of Advanced Computing Symposium, and National Communication Association 2002 Summer Conference, pp. 175–7, May 16–19.

Lazaric, N., Longhi, C. and Thomas, C. (2004) Codification of Knowledge Inside a Cluster: The Case of the Telecom Valley in Sophia Antipolis. Available at www.druid.dk/ocs/viewpaper.php?id=223&cf1 (accessed August 28, 2005).

Leman Group (eds) (1976) *Great Lakes Megalopolis: From Civilization to Ecumenization*, Ottawa: Minister of Supply and Services Canada.

Lilienthal, D.E. (1944) *TVA: Democracy on the March*, New York: Pocket Books.

Lim, G.C. (2003) "Asian Thoughts for Good Governance," in *Life-Culture and Regional Development Planning: Making Gyeonggi a Livable Place*, World Life-Culture Forum 2003, Suwon, Korea: Kyonggi Research Institute, pp. 3–31.

Lo, F.C. and Marcotullio, P.J. (2000) "Globalization and Urban Transformations in the Asia-Pacific Region: A Review," *Urban Studies*, 37 (1): 77–111.

London Development Agency (n.d.). Available at www.lda.gov.uk (accessed June 3, 2005).

Lorentzon, S. (2005) "The Creation of Gothia Science Park in Skövde, Sweden," paper presented at Digital Communities Conference, Benevento, Italy, June.
Lundvall, B.Å. and Johnson, B. (1994) "The Learning Economy," *Journal of Industry Studies*, 1 (2): 23–42.
McNeill, D. (2004) *New Europe: Imagined Spaces*, London: Arnold.
Madanipour, A., Hull, A. and Healey, P. (eds) (2001) *The Governance of Place: Space and Planning Processes*, Aldershot: Ashgate Publishing.
Malecki, E.J. (1997) *Technology and Economic Development* (2nd edn), Harlow: Addison Wesley Longman.
Mann, C.L. (2004) "Information Technologies and International Development: Conceptual Clarity in the Search for Commonality and Diversity," *Information Technologies and International Development*, 1 (2) Winter: 67–79.
Markoff, J. and Richtel, M. (2005) "Profit, Not Jobs, in Silicon Valley," *The New York Times*: 1 and 13, July 3.
Markusen, A. (2000) "Planning As Craft and As Philosophy," in L. Rodwin and B. Sanyal (eds) *The Profession of City Planning: Changes, Images, and Challenges 1950–2000*, New Brunswick, NJ: Center for Urban Policy Research, Rutgers, The State University of New Jersey, pp. 261–74.
—— (2004) "Targeting Occupations in Regional and Community Economic Development," *Journal of the American Planning Association*, 70 (3): 253–68.
Markusen, A. and King, D. (2003) *The Artistic Dividend: The Arts' Hidden Contributions to Regional Development*, Minneapolis, MN: Project on Regional and Industrial Economics, Humphrey Institute of Public Affairs, University of Minnesota.
Markusen, A., Hall, P. and Glasmeier, A. (1986) *High Tech America: The What, How, Where, and Why of the Sunrise Industries*, Boston, MA: Allen & Unwin.
Markusen, A., Lee, Y.S. and DiGiovanna, S. (eds) (1999) *Second Tier Cities: Rapid Growth Beyond the Metropolis*, Minneapolis, MN and London: University of Minnesota Press.
Markusen, A., Schrock, G. and Cameron, M. (2004) "The Artistic Dividend Revisited," Minneapolis, MN: Project on Regional and Industrial Economics, Humphrey Institute of Public Affairs, University of Minnesota.
Marrow, A.J. (1969) *The Practical Theorist: The Life and Work of Kurt Lewin*, New York and London: Basic Books.
Maskell, P. and Kebir, L. (2005) "What Qualifies as a Cluster Theory?" DRUID Working Papers, No. 05–09, Frederiksberg, Denmark: Department of Industrial Economics and Strategy, Copenhagen Business School. Available at www.druid.dk/wp/pdf_files/05-09.pdf (accessed September 19, 2005).
Mayer, H. (2002a) "Taking Root in the Silicon Forest: A Case Study of the Role of Firms as 'Surrogate Universities' in the High Technology Industry in Portland, Oregon," paper prepared for presentation at the 41st Annual Meeting of the Western Regional Science Association, Monterey, California, February 17–20.
—— (May 2002b) "The Silicon Forest Family Tree is Growing," *The Cursor* (a poster). Available at www.upa.pdx.edu/IMS/currentprojects/502SAO%20article.pdf (accessed June 19, 2005).
—— (2003) "A Clarification of the Role of the University in Economic Development," paper presented at the Joint Conference of the Association of the Collegiate Schools of Planning and the Association of European Schools of Planning, Leuven, Belgium, July 8–13.
Mayer, H. and Provo, J. (2004) "The Portland Edge in Context," in C.P. Ozawa (ed.) *The Portland Edge*, Washington, DC: Island Press, pp. 9–34.
Mayer, H., Provo, J. and Seltzer, E. (2004) "New Economy, Old Strategy: Finding a Place for Regionalism in Economic Development," paper presented at the Annual Conference of the Association of Collegiate Schools of Planning, Portland, OR, October 21–24.
Michigancoolcities.com (n.d.) *Michigan Cool Cities Survey*. Available at www.michigan-coolcities.com (accessed June 22, 2005).

Michigan Economic Development Corporation (n.d.) "Michigan Smart Zones." Available at medc.michigan.org/smartzones/index.asp (accessed July 10, 2005).
Ministry of Economic Development (n.d.) "What Is the Knowledge Economy?" Available at www.med.govt.nz/pbt/infotech/knowledge_economy/knowledge_economy-04.html (accessed July 8, 2005).
Ministry of Education, Culture, Sports, Science and Technology (2005) "Major Policies (Science and Technology Policies)." Available at www.mext.go.jp/english/org/eshisaku/ekagaku.htm (accessed July 8, 2005).
Ministry of Information, Communications and the Arts (2002) *create.connect@sg: Arts, Media and Infocomm in Singapore*, Singapore: Ministry of Information, Communications and the Arts.
MIT Industrial Performance Center (n.d.) "Local Innovation Systems Project." Available at ipc-lis.mit.edu/intellectual.html (accessed April 1, 2005).
Mitchell, W.J. (1995) *City of Bits: Space, Place and the Infobahn*, Cambridge, MA: MIT Press.
—— (1999) *e-topia: "Urban Life, Jim—But Not As We Know It,"* Cambridge, MA: MIT Press.
Mitchell, W.J., Inouye, A.S. and Blumenthal, M.S. (2003) *Beyond Productivity: Information Technology, Innovation, and Creativity*, Washington, DC: National Academies Press.
Mokyr, J. (2002) *The Gifts of Athena: Historical Origins of the Knowledge Economy*, Princeton, NJ: Princeton University Press.
Moller, J.O. (1995) *The Future European Model: Economic Internationalization and Cultural Decentralization*, Westport, CT: Praeger.
Moon, Y. (May 2005) "Break Free from the Product Life Cycle," *Harvard Business Review*, 83 (5): 86–94.
Morgan, K. and Nauwelaers, C. (eds) (2003) *Regional Innovation Strategies: The Challenge for Less-Favored Regions*, London and New York: Routledge.
Morrison, S. (2005) "High-Speed Internet Access for All? Not So Fast," *Financial Times*: 9, June 3.
Morris-Suzuki, T. (1994) *The Technological Transformation of Japan: From the Seventeenth to the Twenty-first Century*, Cambridge: Cambridge University Press.
Morris-Suzuki, T. and Rimmer, P.J. (2000) "Cyberstructure and Social Forces – The Japanese Experience," in M.I. Wilson and K.E. Corey (eds) *Information Tectonics: Space, Place and Technology in an Electronic Age*, Chichester and New York: John Wiley & Sons, pp. 117–134.
Moyers, B. (2005) *Moyers on America: A Journalist and His Times*, New York: Anchor Books.
Mucerino, S. and Paradiso, M. (2005) "Mediterranean Agency Remote Sensing – MARS: A Case of Integration of Engineering, Place Sciences and Information Society Policies," paper presented at Digital Communities Conference, Benevento, Italy, June.
Munnich, Jr, L.W., Schrock, G. and Cook, K. (2002) *Rural Knowledge Clusters: The Challenge of Rural Economic Prosperity*, Washington, DC: US Economic Development Administration.
Muscara, C. (ed.) (2003a) "Part 1: Reflections on Gottmann's Thought," *Ekistics*, 70 (418/419): 1–128, Jan./Feb.–March/April.
—— (ed.) (2003b) "Part 2: From Megalopolis to Global Cities," *Ekistics*, 70 (420/421): 129–256, May/June–July/August.
—— (ed.) (2003c) "Part 3: The Bag of Tools for a New Geopolitics of the World," *Ekistics*, 70 (422/423): 257–404, Sept./Oct.–Nov./Dec.
Myers, D. and Banerjee, T. (2005) "Toward Greater Heights for Planning: Reconciling the Differences Between Profession, Practice and Academic Field," *Journal of the American Planning Association*, 71 (2): 121–9, Spring.
Nair, M. (2004) "Competitiveness in the New Economy: New Paradigm and Strategies for Developing Countries," paper presented at the Annual Conference of the Asian Media Information and Communication Centre, Bangkok, Thailand, July.

National Endowment for Science, Technology and the Arts. Available at nesta.org.uk (accessed June 3, 2005).
National Innovation System (n.d.). Available at http://www.absoluteastronomy.com/encyclopedia/N/Na/National_innovation_system.htm (accessed February 23, 2005). Click on OECD publication National Innovation System 1997.
National Research Council (1999) *The Changing Nature of Work: Implications for Occupational Analysis*, Washington DC: National Academy of Science.
Negroponte, N. (1995) *Being Digital*, New York: Vintage Books.
Nora, S., and Minc, A. (1978) *The Computerization of Society: A Report to the President of France*, Cambridge, MA and London: The MIT Press.
Oakland County, Michigan (n.d.) "Wireless Oakland." Available at www.co.oakland.mi.us/wireless (accessed June 25, 2005).
Ogilvy, J.A. (2002) *Creating Better Futures: Scenario Planning as a Tool for a Better Tomorrow*, Oxford: Oxford University Press.
O'Mara, M.P. (2005) *Cities of Knowledge: Cold War Science and the Search for the Next Silicon Valley*, Princeton, NJ and Oxford: Princeton University Press.
Ontario Network on the Regional Innovation System (2005) "Ontario's Regional Economic Development and Innovation Newsletter." Available at www.utoronto.ca/onris/newsletter/0506/newsletterlink109.htm (accessed July 2, 2005).
Organisation for Economic Co-operation and Development (1997) *National Innovation Systems*, Paris: Organisation for Economic Co-operation and Development. Available at www.absoluteastronomy.com/encyclopedia/N/Na/National_innovation_system.htm (accessed July 8, 2005).
—— (2001) *Science, Technology and Industry Outlook, Drivers of Growth: Information Technology, Innovation and Entrepreneurship*, Special Edition 2001, Paris: Organisation for Economic Co-operation and Development.
—— (2003) *Science, Technology and Industry Scoreboard 2003 – Towards a Knowledge-Based Economy*. Available at www1.oecd.org/publications/e-book/92-2003-04-1-7294 (accessed August 28, 2005).
—— (2005) "OECD Key ICT Indicators." Available at www.oecd.org/documentprint/0,2744,en_2649_37409_33987543_1_1_1_37409,00.htm (accessed July 8, 2005).
Ozawa, C.P. (ed.) (2004) *The Portland Edge: Challenges and Successes in Growing Communities*, Washington, DC: Island Press.
Park, W.S. (2003) "Emerging Necessity of Economic Ties between World Class Cities in the Northeast Asian Region," presentation made at the CASID-WID Friday Forum, Michigan State University, April 25.
Pastor Jr, M., Dreier, P., Grigsby III, J.E. and Lopez-Garza, M. (2000) *Regions That Work: How Cities and Suburbs Can Grow Together*, Minneapolis, MN and London: University of Minnesota Press.
Paul, W.G. (1997) "The Electronic Charrette." *Online Planning Journal*. Available at www.casa.ucl.ac.uk/planning/articles11/ec.htm (accessed July 5, 2005).
Peattie, L.R. (1970) "Drama and Advocacy," *Journal of the American Institute of Planners*, 36: 405–10, November.
Peirce, N. (2005) "Katrina's Opportunity: A New New Federalism," *Stateline.org*, September 20. Available at www.stateline.org/live/ViewPage.action?siteNodeId=136&languageId=1&contentId=55197 (accessed September 26, 2005).
Phillips, R.G. (2003) *Evaluating Technology-Based Economic Development: Gauging the Impact of Publicly-Funded Programs and Policies*, Lewiston, NY and Queenston, Ontario and Lampeter, Wales: The Edwin Mellen Press.
Pile, S. and Thrift, N. (eds) (2000) *City A–Z*, London and New York: Routledge.
Plummer, P. and Taylor, M. (2001a) "Theories of Local Economic Growth (part 1): Concepts, Models and Measurement," *Environment and Planning A*, 33: 219–36.
—— (2001b) "Theories of Local Economic Growth (part 2): Model Specification and Empirical Validation," *Environment and Planning A*, 33: 385–98.

—— (2003) "Theory and Praxis in Economic Geography: 'Enterprising' and Local Growth in a Global Economy," *Environment and Planning C*, 21: 633–49.

Policy One Research (2004) "Maine Innovation Index 2004," Portland, Maine: Maine Department of Economic and Community Development.

Porter, M.E. (1990) *The Competitive Advantage of Nations*, New York: The Free Press.

—— (1998) "Clusters and the New Economics of Competition," *Harvard Business Review*, Reprint 98609, Boston, MA: Harvard Business Review: 77–90.

Practice in Planning Education Committee (2004) "Planning Practice Report," draft report, Tallahassee: Association of Collegiate Schools of Planning, July 23. Available at www.acsp.org/practice/default.htm (accessed June 29, 2005).

Pricewaterhousecoopers (2001) *Techmap*: *Waterloo Region 2001 Canada's Technology Triangle* (a poster), Waterloo, Ontario: Pricewaterhousecoopers. Available at www.tech-triangle.com/Publications/ResearchReports.cfm (accessed July 4, 2005).

Progressive Policy Institute (2001) *The Metropolitan New Economy Index*. Available at www.neweconomyindex.org/metro/metro_3mb.pdf (accessed August 24, 2005).

—— (2002) *The State New Economy Index*. Available at www.neweconomyindex.org/states/2002/PPI_State_Index_2002.pdf (accessed August 24, 2005).

Provo, J. (2002) "Planning for Regional Economic Development in Oregon: Finding a Place for Equity Issues in Regional Governance," *Critical Planning* (journal of the UCLA Department of Urban Planning), 9: 55–70.

Public Policy Institute of California (n.d.) "PPCI Bulletin." Available at www.ppic.org (accessed July 5, 2005).

Ravenhill, J. (2001) *APEC and the Construction of Pacific Rim Regionalism*, Cambridge: Cambridge University Press.

Reamer, A. (2003) *Technology Transfer and Commercialization*: *Their Role in Economic Development*, Washington, DC: Economic Development Administration.

Reich, R.B. (1991) *The Work of Nations*, New York: Alfred A. Knopf.

Reid, B. and Morrison, T. (1994) *A Star Danced*: *The Story of How Stratford Started the Stratford Festival*, Toronto: Robert Reid.

Reid, T.R. (1999) *Confucius Lives Next Door*: *What Living in the East Teaches Us About Living in the West*, New York: Vintage Books.

—— (2004) *The United States of Europe: The New Superpower and the End of American Supremacy*, New York: The Penguin Press.

Reiner, T.A. (1963) *The Place of the Ideal Community in Urban Planning*, Philadelphia, PA: University of Pennsylvania Press.

Rheingold, H. (1993) *The Virtual Community*, Reading, MA: Addison-Wesley.

Rifkin, J. (2004) *The European Dream*: *How Europe's Vision of the Future is Quietly Eclipsing the American Dream*, New York: Jeremy P. Tarcher/Penguin.

Rimmer, P.J. (1994) "Regional Economic Integration in Pacific Asia," *Environment and Planning A*, 26: 1731–59.

Ringland, G. (2002) *Scenarios in Public Policy*, Chichester: John Wiley & Sons.

Rivera, D. (2005) "A South Waterfront Player Goes North," *The Sunday Oregonian*: B1-B2.

Rodan, G. (2004) "The Coming Challenge to Singapore Inc," *Far Eastern Economic Review*, 168 (1): 51–4.

Rodwin, L. and Sanyal, B. (eds) (2000) *The Profession of City Planning*: *Changes, Images and Challenges*: *1950–2000*, New Brunswick, NJ: Center for Urban Policy Research, Rutgers, The State University of New Jersey.

Rogers, E.M. (1976) "Communication and Development: The Passing of the Dominant Paradigm," *Communication Research*, 3: 213–40.

Rosenberg, D. (2002) *Cloning Silicon Valley: The Next Generation High-Tech Hotspots*, London: Reuters.

Rural Internet Access Authority [Now known as North Carolina's e-NC Authority] (2002) "Broadband Technology Access Overview," Raleigh, NC: Rural Internet Access Authority. Available at www.e-nc.org (accessed July 7, 2005).

Sampson, D.A. (2002) "Role of Entrepreneurship in Economic Development Strategies," speech given at the EDA (Economic Development Administration) Philadelphia Region 2002 Conference, Philadelphia, April 2.

Sanyal, B. (2000) "Planning's Three Challenges," in L. Rodwin and B. Sanyal (eds) *The Profession of City Planning: Changes, Images, and Challenges 1950–2000*, New Brunswick, NJ: Center for Urban Policy Research, Rutgers, The State University of New Jersey, pp. 312–33.

Sassen, S. (1991) *The Global City: New York, London and Tokyo*, Princeton, NJ: Princeton University Press.

—— (1994) *Cities in a World Economy*, Thousand Oaks, London and New Delhi: Pine Forge Press.

Savage, V.R. (2000) "Renaissance City: Crossing Boundaries for the Arts," in *inform. educate.entertain@sg: Arts & Media in Singapore*, Singapore: Ministry of Information and The Arts, pp. 1–27.

Saxby, S. (1990) *The Age of Information*, New York: New York University Press.

Saxenian, A. (1994) *Regional Advantage: Culture and Competition in Silicon Valley and Route 128*, Cambridge, MA and London: Harvard University Press.

Sayer, A. and Walker, R. (1992) *The New Social Economy*, Cambridge: Basil Blackwell.

Schön, D.A. (1983) *The Reflective Practitioner: How Professionals Think in Action*, New York: Basic Books.

Schön, D., Sanyal, B. and Mitchell, W.J. (1999) *High Technology and Low-Income Communities: Prospects for the Positive Use of Advanced Information Technology*, Cambridge, MA: MIT Press.

Schumpeter, J. (1934) *The Theory of Economic Development*, Cambridge, MA: Harvard University Press.

Scott, A.J. (2000) *The Cultural Economy of Cities: Essays on the Geography of Image-Producing Industries*, London: Sage Publications.

—— (ed.) (2001) *Global City-Regions: Trends, Theory, Policy*, Oxford: Oxford University Press.

Seley, J.E. and Wolpert, J. (2002) *New York City's Nonprofit Sector*, New York: Community Studies of New York and the Nonprofit Coordinating Committee of New York.

Servon, L.J. (2002) *Bridging the Digital Divide: Technology, Community, and Public Policy*, Malden, MA: Blackwell Publishing.

Shapiro, C. and Varian, H. (1999) *Information Rules*, Boston, MA: Harvard Business School Press.

Singh, K.R. (2003) "Michigan's Windows to the Global Knowledge Economy: A County and Regional Level Web Site Analysis from an Economic Development Perspective," Community and Economic Development Occasional Papers, Lansing, MI: Community and Economic Development Program, Michigan State University, July. Available at www.msu.edu/user/cua (accessed May 12, 2005).

—— (2004) "Planning for a Biosciences Based Economy: Exploring the Potentiality in the East Central Michigan Planning and Development Region," unpublished master's thesis, Urban and Regional Planning Program, Department of Geography, Michigan State University.

Smart Michigan (n.d.) "Smart Michigan." Available at http://www.smartmichigan.org (accessed July 10, 2005). Click on "Reports", then click on "Presentations".

Sommers, P. (2002) "Why Do High Tech Firms Locate Where They Do?" presentation made at the Sixth Annual Conference of the State Science and Technology Institute, Dearborn, Michigan, October 2.

Sommers, P., *et al.* (2000) "Ten Steps to a High Tech Future: The New Economy in Metropolitan Seattle," Washington, DC: Center on Urban and Regional Policy, The Brookings Institution.

Sorensen, T. (2004) "Nexus Between Research and Policy Development," unpublished briefing notes prepared for presentation to the Australian government.

Southern Growth Policies Board (n.d.) "Southern Growth Policies Board." Available at www.southern.org (accessed July 4, 2005).

Special Investor.com (n.d.) "Zero-Base Budgeting." Available at www.specialinvestor.com/terms/494.html (accessed July 8, 2005).

SRI International and Michigan Economic Development Corporation (2002) *Benchmarks for the Next Michigan: Measuring Our Competitiveness*, Lansing, MI: SRI International and Michigan Economic Development Corporation.

State Science and Technology Institute (2002) "Program for the 6th Annual Conference," Dearborn, MI: State Science and Technology Institute, October.

—— (n.d.) "SSTI Weekly Digest." Available atwww.ssti.org/digest/digform.htm (accessed July 4, 2005).

Stavrou, S. (2003) *Building Learning Regions: An Innovative Concept for Active Employment Policy*, Rhodes: Deputy Director Cedefop European Forum for Local Development and Employment.

Stefik, M. (1999) *The Internet Edge: Social, Technical, and Legal Challenges for a Networked World*, Cambridge, MA: MIT Press.

Steinfield, C. (ed.) (2003) *New Directions in Research on e-Commerce*, West Lafayette, IN: Purdue University Press.

Sull, D.N. (1999) "Why Good Companies Go Bad," *Harvard Business Review*, product number 4320: 1–10, July–August.

—— (2003) *Revival of the Fittest*, Boston, MA: Harvard Business School Press.

SurveyMonkey.com (n.d.) "Welcome to a Revolutionary Tool." Available at surveymonkey.com (accessed July 9, 2005).

Swann, G.M.P. (1999) "The Internet and the Distribution of Economic Activity," in Macdonald, S. and Nightingale, J. (eds) *Information and Organization*, Amsterdam: Elsevier Science B.V., pp. 183–95.

Tatsuno, S. (1986) *The Technopolis Strategy: Japan, High Technology, and the Control of the Twenty-first Century*, New York: Prentice Hall Press.

Taylor, P.J. (n.d.) "Atlas of Hinterworlds." Available at www.lboro.ac.uk/departments/gy/gawc/visual/hwatlas.html (accessed July 4, 2005).

—— (1997) "Hierarchical Tendencies amongst World Cities," *Research Bulletin 1*, published in *Cities*, 14 (6): 323–32. Available at www.lboro.ac.uk/gawc/rb/rb1.html (accessed July 4, 2005).

—— (1999) "Places, Spaces and Macy's: Place-Space Tensions in the Political Geography of Modernities," *Progress in Human Geography*, 23 (1): 7–26.

—— (2001) "Specification of the World City Network," *Geographical Analysis*, 33 (2): 181–94.

—— (2004a) *World City Network: A Global Urban Analysis*, London and New York: Routledge.

—— (2004b) *Understanding London in a New Century*, Globalization and World Cities Study Group and Network Research Bulletin 138. Available at www.lboro.ac.uk/gawc/rb/rb138.html (accessed June 20, 2004).

Technology Alliance (2000) "The Economic Impact of Technology-Based Industries in Washington State." Available at www.technology-alliance.com/pubspols/studies/techecon_2000_execsum.html (accessed July 8, 2005).

Tennessee Valley Authority (n.d.) "Tennessee Valley Authority: Energy, Environment, Economic Development." Available at www.tva.gov (accessed July 5, 2005).

Thant, M. and Tang, M. (eds) (1996) *Indonesia–Malaysia–Thailand Growth Triangle: Theory to Practice*, Manila: Asian Development Bank.

Thrush, G. (n.d.) "Ring City." Available at projects.gsd.harvard.edu/appendx/dev/issue3/thrush/index1.htm (accessed October 16, 2003).

Toffler, A. (1970) *The Third Wave*, New York: Bantam Books.

Tornatzky, L.G., Waugman, P.G. and Gray, D.O. (2002) *Innovation U.: New University Roles in a Knowledge Economy*, Research Triangle Park, NC: Southern Growth Policies Board.

Turkle, S. (1995) *Life on the Screen*, New York: Simon & Schuster.

United Nations Development Programme (2004) *Sources for Democratic Governance Indicators*, New York: United Nations Development Programme. Available at www.grc-exchange.org/info_data/record.cfm?Id=1124 (accessed June 28, 2005).

United Nations Economic and Social Commission for Asia and the Pacific (2002) "What is Good Governance?" Available at www.unescap.org/huset/gg/governance.htm (accessed July 8, 2005).

United States Census (2001) *Statistical Abstract of the United States*, Washington, DC: USGPO.

United States Department of Commerce (2002) "Instructions for Survey of Industrial Research and Development During 2001," Form RD-1A. Available at www.nsf.gov/sbe/srs/sird/form2001/rd1a.pdf (accessed July 8, 2005).

United States Economic Development Administration (n.d.) "Economic Development Administration Home." Available at www.osec.doc.gov/eda/html/2al_whatised.htm (accessed July 7, 2005).

University of Arizona (n.d.) "Course for Studying the Future." Available at ag.arizona.edu/futures/home/glossary.html (accessed April 7, 2005).

Urban Institute Ireland (n.d.) POLYNET – Sustainable Management of European Polycentric Mega-City Regions. Available at www.urbaninstitute.net/research_polynet.shtml/ (accessed May 8, 2005).

Urban Redevelopment Authority (n.d.) "Creating a More Beautiful Singapore," Singapore: Urban Redevelopment Authority.

Van de Ven, A.H. and Koenig, Jr, R. (1976) "A Process Model for Program Planning and Evaluation," *Journal of Economics and Business*, 23 (3): 161–70, Spring–Summer.

Van Looy, B., Debackere, K. and Andries, P. (2003) "Policies to Stimulate Regional Innovation Capabilities Via University-Industry Collaboration: An Analysis and an Assessment," *R & D Management*, 33 (2): 209–29.

Virtual Charlottetown (n.d.) "Charlottetown Set to Become One of Canada's First e-Cities: P.E.I. Capital Launched as Virtual City and Area Residents will Soon be Conducting Their Business Online." Available at www.virtualcharlottetown.com/newsresult.cfm?ID=19 (accessed November 19, 2001).

Vision 2010 (n.d.) "Vision 2010: Universities in the 21st Century." Available at www.si.umich.edu/V2010/home.html (accessed July 8, 2005).

Warf, B. (1989) "Telecommunications and the Globalization of Financial Services," *Professional Geographer*, 41 (3): 257–71.

Webber, M.M. (1968) "The Post-City Age," *Daedalus, Journal of the American Academy of Arts and Sciences*, 97 (4): 1091–110, Fall.

Williams, F. (2003) "Asia Nations Break into Top League for Internet Access," *Financial Times*: 6, November 20.

Williams, J.F. and Stimson, R.J. (eds) (2001) *International Urban Planning Settings: Lessons of Success*, International Review of Comparative Public Policy, Vol. 12, Amsterdam: JAI and Elsevier Science.

Wilson, M. I. (1995) "The Office Farther Back: Business Services, Productivity, and the Offshore Back Office," in P. Harker (ed.) *Service Management, Technology and Economics: The Service Productivity and Quality Challenge*, Dordrecht: Kluwer, pp. 203–24.

—— (1998) "Information Networks: The Global Offshore Labor Force," in G. Sussman and J.A. Lent (eds) *Global Production: Labor in the Making of the 'Information Society'*," Cresskill, NJ: Hampton Press.

Wilson, M.I. and Corey, K.E. (eds) (2000) *Information Tectonics: Space, Place and Technology in an Information Age*, Chichester: John Wiley & Sons.
Wilson, M.I., Corey, K.E., Helmholdt, N. and Frederick, E. (2004) "Technology, Social Capital and Urban Life in Michigan: Informing the Debate – The Built Environment, Social Capital, and Human Health," report submitted to the Michigan Applied Public Policy Research Program of the Institute for Public Policy and Social Research, Michigan State University. Available at www.smartmichigan.org; and www.ssc.msu.edu/~espace/MichiganInternet2004pdf (accessed July 6, 2005).
Wolfe, D.A. (ed.) (2003) *Clusters Old and New: The Transition to a Knowledge Economy in Canada's Regions*, Montreal and Kingston: McGill-Queens University Press.
—— (2004) "The Dynamics of Cluster Development," presentation to the Cluster Strategy Implementation Workshop, Edmonton, January 22.
Wolfe, D.A. and Gertler, M.S. (n.d.) "Innovation Systems and Economic Development: The Role of Local and Regional Clusters in Canada," Toronto: Program on Globalization and regional Innovation Systems, University of Toronto.
Wolfe, D.A. and Lucas, M. (eds) (2004) *Clusters in a Cold Climate: Innovation Dynamics in a Diverse Economy*, Montreal and Kingston: McGill-Queens University Press.
World Bank Group (2001) "Indicators of Governance and Institutional Quality." Available at www1.worldbank.org/publicsector/indicators.htm (accessed June 28, 2005).
Yeung, H.W.C. (1998) "Capital, State and Space: Contesting the Borderless World," *Transactions of the Institute of British Geographers*, 23: 291–309.
—— (2002) "Towards a Relational Economic Geography: Old Wine in New Bottles?" paper presented at the 98th Annual Meeting of the Association of American Geographers, Los Angeles, March 13.
—— (2003) "Practicing New Economic Geographies: A Methodological Examination," *Annals of the Association of American Geographers*, 93 (2): 442–62.
—— (2005) "Rethinking Relational Economic Geography," *Transactions of the Institute of British Geographers*, 30: 37–51.
Yeung, H.W.C. and Lin, G.C.S. (2003) "Theorizing Economic Geographies of Asia," *Economic Geography*, 79 (2): 107–28, April.
Yu, W.I. (1996) "A New Capital for Unified Korea," *Korea Focus*, 4 (1): 53–61, January–February.
Zhang, J. and Patel, N. (2005) *The Dynamics of California's Biotechnology Industry*, San Francisco, CA: Public Policy Institute of California.
Zook, M.A. (2005) *The Geography of the Internet Industry: Venture Capital, Dot-Coms, and Local Knowledge*, Malden, MA: Blackwell Publishing.

Index

Note: Page references in *italics* indicate illustrations.
The abbreviation ICT stands for information and communications technologies.